D1556452
ERAU HAZY LIBRARY

LUFTWAFFE RUDDER MARKINGS 1936-1945

Karl Ries/Ernst Obermaier

LUFTWAFFE RUDDER MARKINGS 1936-1945

SCHIFFER MILITARY HISTORY
West Chester, PA

Library of Congress Catalog Number: 91-61684.

Printed in the United States of America.
ISBN: 0-88740-337-9

This title was originally published under the title,
Bilanz am Seiten-leitwerk,
by Verlag Dieter Hoffmann, Mainz.

Published by Schiffer Publishing, Ltd.
1469 Morstein Road
West Chester, Pennsylvania 19380
Please write for a free catalog.
This book may be purchased from the publisher.
Please include $2.00 postage.
Try your bookstore first.

Introduction

A theme somewhat tied-in with the Hitler era in Germany should be – according to a widespread belief among German journalists – treated only with extreme delicacy and always with a certain feeling of guilt. Everything that is connected with the years 1933 to 1945 is to this day, 25 years afterwards, burdened with some journalistic taboos that has may people shaking their heads. Apparently the fact is overlooked that this period is a part of history, the same as numerous others tens or hundreds of years ago, either more or less successful, something to be either proud or ashamed of. Certainly people of various denominations, even those completely unprejudiced, will respond to such a book with a wide range of reactions: the conscientious pacifist will not buy it in the first place, which is sensible since the book does not offer him anything; the diehard traditionalist will perhaps place it front center in his bookcase – after all, there are lots of swastikas in it; the well-to-do hippie may utilize it in the course of a happening (caution, stiff paper); but the aviation enthusiast will regard it as a reminder of a certain period in history that had been neglected too long.

Military historians as well as young or »ancient« flyers will find on the following pages some data that, in their compactness and their details, can hardly be found in any other publication. The old »Pilotiseur« who used to have some »kill bars« on his rudder may, when viewing these pictures, still hear the ugly sound on his aircraft getting hit. Many, very many pilots, German and – then – enemy ones whose luck did not hold out cannot turn the pages of this book anymore and particularly to them, who helped to write this part of history, this book is dedicated.

Once the war machinery had been started it did not make the slightest difference whether a pilot did his duty out of some sort of conviction or merely »pushed the button« because he was simply frightened. The moment of thruth left only one alternative, (pardon the banality) You or Me! This consideration should help in solving the apparent riddle whether the participants in this drama were cold-blooded killers, cowards, heroes or simply ambitious fools. A »matter of routine«, as called so often, the downing of an enemy never became. Only his ability of handling his aircraft and his flying skill had a pilot achieve his successes. Often enough, this should not be overlooked either, it was the faithful wing man too who did the »covering«, often by disregarding his own safety.

A chain of philosophical and psychological thoughts could be spun on the theme: thoughts upon shooting down or being shot down. But his cannot be the task of this book. Its intention is to give a compilation of special markings, the success markings on aircraft of the German Luftwaffe, based on documentary proof.

The team of authors could never have offered such exactly dated pictures and victories if it had not been for Hans Ring's incomparable collection of aerial victory reports. To him we are especially grateful and also to the pilots or relatives of deceased Luftwaffe members who made available picture and documentary material.

Finthen, 20. 8. 1970

Karl Ries

Munich, 20. 8. 1970

Ernst Obermaier

CONTENTS

1. From »Corn Plaster« to »Kill Bar«.

As our predecessors from the paleolithic age had a functional use for their animal skins, as cover against the weather and the cold, the very first »Flying Machines« made an equally functional use of their fabric covering in keeping the whole flimsy contraption together and flying. These ancient aeroplanes prior to the year 1910 with their sometimes baffling composition of struts, wires and fabric were the very picture of a strictly technical device and, in most cases, not a very aesthetic one at that.

It did not take long, however, until some bright person had the idea to work on these many square feet of canvas with paint and brushes, possibly with the purpose of tidying up the ugly looks of that stringbag but in most cases with the intention of using the hitherto blank space for advertising. In the beginning there were merely the names of the designers or builders painted on, to be followed only a short time later by some inscriptions that were outside of the aviation scope, an example were the »Batschari« cigarette advertisements on aircraft of that period. Even then these advertising offers were welcomed by the aircraft owners with open arms since they helped to easen the financial burden of their costly sport.

With the outbreak of World War 1 all civilian flying activities came to a complete halt, at least in Germany. All private aircraft were taken over by the Air Arm and received as the national emblem the iron cross in its many variants. Strict orders not to apply anything but national emblem and type designation were disregarded only following the first aerial successes in the year 1915. At this time the first personal emblems made their appearance on aircraft, some of them being of the figurative kind, others were based on geometric patterns that gave the aircraft a rather curious look. Compared to the Allied flying units the German Air Arm had only little in the way of wing or squadron symbols whereas the strictly personal emblems were found in much greater numbers and were reaching an all-time high towards the end of the war.

Despite considerable successes of individual pilots of the fighting nations it did, however, not become customary to mark the number of shot-down enemies on one's own aircraft. The authors know of no single instance du-

1. Vom »Hühneraugenpflaster« zum Abschußbalken

Wie unseren Vorfahren des Paläolithikum ihr Tierfell rein funktionell als Bespannstoff – hier gegen Einflüsse von Witterung und Temperatur – diente, so hatte die Leinwand als Bespannstoff der ersten Fluggestühle die einzige Aufgabe das Gerippe der klapprigen Aeroplane zusammen, und in der Luft zu halten. Die Fluggeräte der aeronautischen »Urzeit« vor 1910 boten somit in ihrer Komposition von Streben, 'erspanndrähten und Stoffbespannnung dem Betrachter das Bild eines rein technischen Gerätes, und in den meisten Fällen nicht einmal ein sehr ästethisches.

Es dauerte jedoch gar nicht allzu lange bis ein findiger Kopf auf den Gedanken kam die vielen Quadratmeter Bespannnung mit Farbe und Pinsel zu bearbeiten, möglicherweise um den schaurigschönen Anblick einer »Drahtkomode« etwas zu mildern, meist jedoch in der Absicht die nackten Flächen zu Reklamezwecken heranzuziehen. Wurden anfangs nur die Namen von Erbauer oder Konstrukteur eines Flugzeuges auf Rumpf und Tragflächen gemalt, so erscheinen wenig später die ersten spartenfremden Werbesprüche, z.B. der Zigarettenfabrik »Batschari« auf den Konstruktionen dieser Jahre. Schon damals kamen den Flugzeugbesitzern Werbe-Angebote dieser Art sehr gelegen, halfen sie doch mit, den recht kostspieligen Sport des Fliegens zu finanzieren.
Mit Ausbruch des 1. Weltkrieges kommt, in Deutschland zumindest, die zivile Fliegerei vollkommen zum Erliegen. Alles private Flugzeuggerät wird von der Fliegertruppe übernommen, und erhält als Nationalitäts-Kennzeichen das Eiserne Kreuz in seinen vielen Varianten. Die strikte Anordnung, den Flugzeugen außer Nationalitätszeichen und Typenbezeichnung keine weitere Bemalung zu geben, wird erst nach den ersten fliegerischen Erfolgen des Jahres 1915 mißachtet. Zu diesem Zeitpunkt erscheinen die ersten persönlichen Kennzeichen auf den Flugzeugen, wobei solche Markierungen teils figürlicher Art sind, teils geometrische Muster zugrunde liegen, die mitunter den Maschinen ein skurriles Aussehen verleihen.

In geringerem Maße als bei den alliierten Fliegerstaffeln zeigen sich bei der Deutschen Fliegertruppe Geschwader- und Staffelzeichen; dagegen wird, wie bereits vorbemerkt die Anbringung eines persönlichen Kennzeichens weitaus häufiger praktiziert, und erreicht gegen Kriegsende auf deutscher Seite einen schier unübersehbaren Formenreichtum.

Trotz der beachtlichen Abschußerfolge einzelner Piloten der kriegsführenden Nationen bürgert es sich während des 1. Weltkrieges dagegen nicht ein, die Zahl der erfolgreich bestandenen Luftkämpfe mit Abschuß eines Gegners auf der eigenen Maschine

ring World War 1 where a German fighter pilot bore such a score on his aircraft. Curiously enough the battle scars received by one's own aircraft, either from a dog fight or anti-aircraft guns, are especially marked. After returning from a mission crew and maintenance personnel eagerly went to work counting enemy hits which, in most cases, merely went right through fabric or plywood covering without doing much damage. Each single bullet hole was repaired, the attacker's national emblem, the roundel, was painted around it and the date of the shooting was added. As long as aerial fighting was done with the help of rifles, pistols and single machine guns these bullet scars could well be counted and the sport of sticking »corn plasters«, as it was then called, was quite popular. But with the increase in number and rate of fire of the machine guns used it became no longer practical to repair single holes, instead a complete new covering of the damaged aircraft portions became a necessity. As a result of this the »corn plaster stickers« in the West stopped their activities very soon, although in some of the more »quiet« theaters of war like Galicia and Italy this custom may have lingered on for some time. Anyway, towards the end of the war in 1918 very few markings of this kind were still to be found among the often replaced and used-up equipment of the German Air Arm.

After the cessation of the hostilities all aviation in Germany was restricted to the civilian sector for a considerable period of time as a result of the Versailles Treaty. Although there was some clandestine rebuilding activity with the aim of resurrecting a new Air Arm but even the small number of Reichswehr aircraft were kept away from the local population's eyes. Not until the year 1935 the new Luftwaffe stepped out of its hidings and became the third, if still modest, part of the Wehrmacht alongside the Heer (Army) and the Kriegsmarine (Navy).

Only slowly the German aircraft industry succeeded in closing the technical gap to foreign aviation development. Forming of the flying units too war taking considerably longer than called for in the plans, even by using the cell-dividing method.

In order to obtain as much organisational and technical know-how in the field of modern air

zu markieren. Den Verfassern ist kein Fall bekannt, wo ein deutscher Jagdflieger bis Ende des Krieges eine solche Erfolgsbuchhaltung auf seinem Flugzeug geführt hätte.

Verblüffenderweise werden im Gegenteil die Narben, die Flakbeschuß oder Luftkampf in der eigenen »Kiste« hinterlassen haben besonders gekennzeichnet. Nach Rückkehr von einem Frontflug machten sich Besatzung und Monteure über den angeschossenen Vogel her, um die Einschüsse zu zählen, die meist nur Bespannstoff oder Sperrholzbeplankung durchschlugen, ohne größeren Schaden anzurichten. Jedes Einschußloch wurde fein säuberlich verklebt, und um die Einschußstelle die Kokarde des Angreifers gemalt, dazu das Datum des überstandenen Beschusses gesetzt. Solange sich die Luftkämpfe noch mit Schnelladegewehren, Pistolen und nur mit einem MG bewaffneten Flugzeugen abspielten, blieben die Einschußstellen übersehbar, und der Sport »Hühneraugenpflaster« zu kleben, wie man es damals bezeichnete, wurde mit Eifer betrieben. Mit Steigerung der eingebauten Waffenzahl und der Feuerfolge der verwendeten MGs, lohnte es sich jedoch bald nicht mehr die Löcher und Risse in der Bespannung zu verkleben, vielmehr wird bei solch arg zerrupften »Heimkehrern« eine Neubespannung der beschädigten Flugzeugteile erforderlich. So stellten die »Hühneraugenpflasterkleber« an der Westfront sehr bald ihre Tätigkeit ein, an ruhigeren Fronten wie Galizien, Italien etc. mag sich dieser Brauch etwas länger gehalten haben. Jedenfalls sind gegen Kriegsende 1918 unter dem häufig wechselnden und schnell verbrauchten Flugzeugmaterial der Deutschen Fliegertruppe kaum noch Markierungen dieser Art zu finden.

Nach Abschluß der Feindseligkeiten bleibt die Fliegerei in Deutschland infolge der Bedingungen im Versailler Vertragswerk für lange Jahre auf den zivilen Sektor beschränkt. Zwar wird im geheimen am Aufbau einer neuen militärischen Fliegertruppe gearbeitet, jedoch bleiben die schwachen Kontingente der Reichswehr-Luftwaffe selbst dem Einheimischen verborgen. Erst mit dem Jahre 1935 tritt die Luftwaffe aus ihrem bisherigen Untergrunddasein an die Öffentlichkeit, und damit als ein noch schwacher Wehrmachtsteil neben die bereits bestehenden, das Heer und die Kriegsmarine.

Nur langsam gelingt es der deutschen Flugzeugindustrie den technischen Vorsprung ausländischer Flugzeughersteller aufzuholen. Auch der Aufbau weiterer fliegender Verbände nach dem System der Zellenteilung bleibt hinter den Aufstellungsplänen des Luftfahrtministeriums erheblich zurück.

Um diesen Rückstand an organisatorischer und technischer Luftkriegserfahrung, der gegenüber anderen Luftmächten besteht auszugleichen, nimmt

DFW C V mit »Hühneraugenpflastern« auf Höhen- und Seitenleitwerk, sowie beiden Tragflächen. 1917, Westfront.

A DFW C V with »Corn Plasters« on empennage and wings. Western front 1917.

Bf 109 B-2 der J 88 »Legion Condor« im Nov. 1937 in Burgos/Spanien. Das Flugzeug trägt einen Abschußbalken am Seitenleitwerk.

A Bf 109 B-2 of the J 88 »Legion Condor« in Burgos, Spain, in November 1937. The aircraft bears one »kill« bar on the vertical stabilizer.

war as possible the German government was only too willing to respond favorably to a call for help by the Spanish National Forces to assist them in their fight against the communist-socialist People's Front. At first small advanced units were sent to Spain to make the necessary preparations prior to the arrival of the bulk of the »Legion Condor« in late 1936. The Legion's flying units comprised the A 88 (Aufklaerungsgruppe – Reconnaissance Group), AS 88 (Aufklaerungs- und Bomberstaffel See – Reconnaissance and Bomber Squadron Sea), J 88 (Jagdgruppe – Fighter Group), and K 88 (Kampfgruppe – Bomber Group). The flying equipment corresponded to the one used by the Luftwaffe units back home. The pilots of the »Legion Condor« found out very soon that their opponents, coming from various nations, were to be taken very seriously and only by use of more modern aircraft as well as by thorough evaluation of the experience gained in the first few months of fighting in Spain were the Legion's pilots able to gain a marked air superiority.

Being equipped with new aircraft of the types Bf 109B, C and E instead of the old He 51's the J 88's pilots achieved their first series of »kills« which were now marked conspicuously on their aircraft for the first time. A rule developed by which aerial victories were marked by vertical bars on the empennage, on both sides of the vertical stabilizer. Each »kill«, one- or multi-engined, was marked by a white bar painted on the grey camouflage paint of the fighter's tail surface. Date of victory and nationality of the enemy were not marked especially with this kind of bars during the Spanish Civil War. After the end of the hostilities in the spring of 1939 a number of German fighter pilots had reached a considerable number of »kills« - Moelders 14, Schellmann 12, Seiler 9, Bertram 9, Balthasar 7, Ihlefeld 7, Tietzen 7. But the RLM, the German Air Ministry, forbade the returning fighter pilots to keep these bars on the aircraft flown with their units in Germany. Even after the outbreak of World. War 2 it was not allowed to add the Spanish victories to the ones gained after 1 September 1939.

As a result of the combined efforts of the German aviation industry and the RLM (Reichsluftfahrtministerium – German Air Ministry) in the years 1935 to 1939 the Luftwaffe

die damalige Deutsche Reichsregierung nur zu gerne die Gelegenheit wahr, einen Hilfsruf der nationalen Kräfte in Spanien nach Unterstützung gegen die kommunistisch-sozialistische Volksfront Folge zu leisten.

Es kommt nach Entsendung kleiner Vorkommandos zur Aufstellung der »Legion Condor«, die mit der Masse ihrer Verbände Ende 1936 auf der Iberischen Halbinsel eintrifft. Die fliegenden Truppenteile der »Legion« setzen sich aus der A 88 (Aufklärungsgruppe), AS 88 (Aufklärungs- und Bomberstaffel See), J 88 (Jagdgruppe) und K 88 (Kampfgruppe) zusammen, wobei das verwendete Fluggerät dem der im Reichsgebiet stationierten Einheiten der Luftwaffe entspricht. Die Flieger der »Legion Condor« bekommen sehr bald zu spüren, daß die auf Seiten der »Rojos« kämpfenden, aus allen möglichen Nationen stammenden Kontrahenten durchaus als ebenbürtige Gegner anzusehen sind; erst durch Umrüstung auf andere Flugzeugmuster, sowie die in den ersten Monaten der Kämpfe in Spanien erlangten taktischen und strategischen Erkenntnisse, gelingt es den fliegenden Teilen der »Legion« eine merkliche Überlegenheit gegenüber der gegnerischen Seite zu erlangen.

Die Neuausstattung der J 88 mit der Bf 109B, C und E, anstelle der bisher geflogenen He 51, bringt den Piloten der Jagdstaffeln die ersten Abschuß-Serien, die nun erstmals augenfällig auf den erfolgreichen Maschinen markiert werden.

Es bildet sich die Regel heraus, die Abschüsse in Form von senkrecht stehenden Balken auf dem Seitenleitwerk, und zwar auf der Seitenflosse beidseitig aufzuzeichnen. Dabei wird je Abschuß – gleichviel ob ein- oder mehrmotorig – ein Abschußbalken in weißer Farbe auf die graue Tarnfarbe der eigenen Jagdmaschine aufgemalt. Das Datum des Abschusses und das Nationalitätszeichen des bezwungenen Gegners werden bei dieser Markierung während des Spanischen Bürgerkrieges nicht besonders vermerkt. Bei Abschluß der Feindseligkeiten im Frühjahr 1939 weisen einige der eingesetzten deutschen Jagdflieger (Mölders 14, Schellmann 12, Seiler 9, Bertram 8, Balthasar 7, Ihlefeld 7, Tietzen 7) bereits eine beachtliche Abschußzahl auf. Den aus Spanien zurückkehrenden Jagdfliegern ist es seitens des RLM untersagt diese Abschußbalken auf den bei ihrem Verband im Reichsgebiet geflogenen Maschinen beizubehalten. Auch nach Beginn des 2. Weltkrieges bleibt es verboten die Spanien-Abschüsse zu den nach dem 1. 9. 1939 errungenen Erfolgen hinzuzusetzen.

Bedingt durch die Anstrengungen, die Flugzeugindustrie und RLM in den Jahren zwischen 1935 und 1939 unternommen hatten, steht die Luftwaffe im

Eine in Frankreich notgelandete Supermarine »Spitfire II«, auf deren Motorenhaube 29 Hakenkreuze als Erfolgszeichen aufgemalt sind.

A Supermarine »Spitfire II«, force-landed in France, on whose engine cowl 29 swastikas have been painted as victory marks.

Der Engländer Clive R. Caldwell, eines der Jagdfliegerasse der Royal Air Force vor seiner »Spitfire« mit deutschen, italienischen und japanischen Siegeszeichen.

Clive R. Caldwell, one of the Royal Air Force's fighter aces, in front of his »Spitfire« bearing German, Italien and Japanese victory marks.

was enjoying both personnel and material superiority over its first opponent, Poland, in September 1939. It was therefore hardly surprising that the aerial victories by Polish pilots bore no comparison with their Luftwaffe opponents. As in the Spanish Civil War the »kills« in Poland were marked on the German aircraft tailplanes without the date of »kill« or the nationality of the enemy, since there was only one enemy at the time. The bars were white and contrasted to the aircraft's camouflage paint (Nr. 70 black-green and Nr. 71 dark green).

After France and Great Britain had joined the fighting and their first successful encounters with aircraft and pilots from these nations the German pilots began adding to their »kill bars« the respective enemy's roundel, sometimes also adding the date. When the conflict widened into a world war the number of participating nations increased and so did the variety of roundels in combination with »kill bars« on the tails of fighter aircraft. Very soon some individual pilots reached such a high score that their vertical stabilizers became too small for this kind of bookkeeping, or the German national emblem, the swastika, could no longer be distinguished from the »kill bars« from some distance. Therefore the still unmarked area of the vertical rudder was used for marking the score. Since the new camouflage paint for fighter aircraft had been introduced, the light blue color Nr. 65 was extended to the fuselage sides and the vertical tailplanes and the »kill bars« were therefore painted in black, sometimes in red.

Pilots with a high »kill« score, when changing over to newer aircraft, often tended to disregard the sequence of their old victories and added the respective nationalities up to rows that were marked by one centrally placed roundel (as with Rudorffer, Gratz, Wurmheller).

When awarded a decoration, from the Ritterkreuz (Knight's Cross) on, most pilots put a symbol of this decoration plus the then achieved number of victories in the upper portion of the tailplane area and placed all the following »kill bars« underneath. Until the end of World War 2 quite a number of variations on this marking theme made their appearance since there never was an official regulation on how to mark one's »kills«. Thus the individual pilot's and/or squadron pain-

September 1939 ihrem ersten Kriegsgegner Polen mit personeller und materieller Überlegenheit gegenüber. Entsprechend diesen Vorteilen deutscherseits ist es nicht verwunderlich, daß die Abschußerfolge der polnischen Jagdstaffeln in keinem Verhältnis zu denen der Luftwaffe stehen. Wie zuvor im Spanischen Bürgerkrieg werden nun auch die Abschüsse während des Polenfeldzuges ohne Angabe des Datums und der Nationalität des abgeschossenen Flugzeuges – schließlich hatte man es nur mit einem, dem polnischen Gegner zu tun – an der Flosse des Seitenleitwerkes markiert, immer noch im Kontrast zur Tarnfarbe des Flugzeuges (Nr. 70 schwarzgrün – Nr. 71 dunkelgrün) in weißer Farbe. Nach Eingreifen der Westmächte, Frankreichs und Englands in das Kriegsgeschehen, und dem ersten Zusammentreffen deutscher Flugzeuge mit solchen dieser Nationen, gehen die erfolgreichen Flugzeugführer dazu über in, oder über dem Abschußbalken die Nationalität des besiegten Gegners in Form der entsprechenden Kokarde anzugeben. Teilweise wird auch bereits das Abschußdatum mit registriert.

Mit der Ausweitung des Konfliktes zu einem Weltkrieg größten Ausmaßes, nimmt naturgemäß die Zahl der beteiligten Nationen und damit die Vielfalt der Kokarden in Verbindung mit Abschußbalken auf den Seitenflossen der Jagdflugzeuge zu. Sehr bald erreichen einzelne Piloten solch hohe Abschußzahlen, daß die Seitenflosse für die Buchhaltung der Abschüsse nicht mehr ausreicht, oder das deutsche Nationalitätszeichen (Hakenkreuz) auf einige Entfernung von den Reihen der Abschußbalken nicht mehr klar zu unterscheiden ist. Es wird daher die freie Fläche des Seitenruders zur Aufzeichnung der Abschüsse benutzt, wobei nunmehr, wie bereits seit Einführung des neuen Tarnanstriches für Jagdflugzeuge (Hochziehen der hellblauen Farbe Nr. 65 über Rumpfseiten und Seitenleitwerk) der Abschußbalken mit schwarzer, oft auch mit roter Farbe aufgetragen wird.

Piloten mit hohen Abschußzahlen lassen beim Umsteigen auf ein Baumuster neuerer Serie dabei sehr oft die Reihenfolge ihrer bisherigen Abschüsse außer Acht, ziehen die einer Nationalität zu einer Balkenreihe zusammen und markieren nur einmal, meist in der Mitte der Reihe die entsprechende Kokarde (s. Rudorffer, Gratz, Wurmheller).

Bei Verleihung einer Auszeichnung ab Ritterkreuz, setzen die meisten Flugzeugführer das Symbol dieser Dekoration mit der bei der Verleihung erzielten Abschußzahl numerisch in den oberen Teil der Ruderfläche, um darunter die folgenden Abschüsse in Form von Balken weiter zu markieren. Es tauchen bis Kriegsende selbstverständlich noch eine ganze Reihe von Markierungsvarianten auf, da ja eine

Auch die Amerikaner vermerkten ihre Bilanz unterhalb der Kabine, wie hier Lt. Col. Robert S. Johnson von der 56th Fighter Group. Am Ende des Krieges hatte Johnson 28 bestätigte Abschüsse.

The Americans too marked their score on the cockpit flanks as in this picture Lt. Col. Robert S. Johnson of the 56th Fighter Group. At the end of the war he had been credited with 28 confirmed victories.

Hakenkreuzflaggen auf der P-51 D des Capt. Ray S. Wetmore, 370th Fighter Squadron US Army Air Force. Wetmore's Abschußkonto wies bei Ende der Feindseligkeiten 22,6 Luftsiege auf.

Swastika flags on the P-51D of Capt. Ray S. Wetmore, 370th Fighter Squadron of the US Army Air Force. At the end of the hostilities he had 22,6 aerial victories to his credit.

ter's phantasy was not limited at all. In some instances it led to the point that several unit leaders had their score put on even their »hack« aircraft's tailplanes.

As the war went on, the flying units were – via the higher echelons in the Luftflotten – repeatedly ordered to discontinue the use of rudder »kill« markings in order to prevent easy recognition of successful fighter pilots by the enemy. The ever increasing growth of the Allied air power resulted in fact in a situation, especially during the last two years of the war, that as soon as a German fighter ace was recognized by his rudder markings or a unit leader by his staff symbols he was attacked and chased by several enemy fighters at once. To avoid such unpleasant encounters that often had fatal results many expert pilots gave up their »calling cards« on the rudder whereas others, be it the result of stubbornness, vanity, self-esteem or the wish to serve as an example for younger pilots, carried on with this practice right to the end of the war. This recognition of the enemy's superiority did not make any difference to those pilots who had never started this practice in the first place, among them well-known aces like Batz, Buehligen and Rall, just to name a few.

It should be mentioned at this point that such methods of keeping score of shot-down enemies during World War 2 were by no means restricted to the Luftwaffe alone but by the Air Forces of all the Allied nations. But whereas German pilots preferred the tailplanes as a place to keep tab the Allied pilots put their »kill« score markings on the fuselage sides underneath the canopy. Only with Germany's Allies, Italy, Hungary, Romania, Croatia, Bulgaria, Finland etc., were the victory marks put on their victorious pilots' rudders. The accompanying pictures are intended to give some idea of these details as practiced by other Air Forces.

Very early in the war Luftwaffe pilots began to differentiate their successes by special symbols. In this way enemy aircraft destroyed on the ground were marked by bars pointed at the lower end (as with Balthasar and Victor Bauer) or even marked on the elevator (as with Hachfeld). Destroyed barrage or observation balloons appeared as silhouettes, the same went for destroyed tanks or

Vorschrift über die Art der Ausführung einer Abschuß-Buchhaltung auf den Maschinen nie erlassen wurde. So bleibt der Phantasie des Flugzeugführers bzw. der Staffelmaler ein weiter Spielraum, der dann auch großzügig ausgenutzt wird, mitunter sogar dahin führt, daß einzelne Verbandsführer ihr Abschußkonto auch auf das Ruder Ihrer Reise- und Kuriermaschine übertragen lassen.

Im Verlauf des Krieges wird mehrfach den Verbänden über die Luftflotten mitgeteilt, daß das Kennzeichnen von Abschüssen auf den Rudern einzustellen ist, um das Erkennen besonders erfolgreicher Jagdflieger dem Gegner zu erschweren. Tatsächlich zeigt es sich mit dem Anwachsen der gegnerischen Luftherrschaft in den letzten beiden Kriegsjahren, daß beim Erkennen eines Jäger-Asses an der Rudermarkierung, oder eines Verbandsführers an seiner Stabs-Kennzeichnung, diese sofort von mehreren Gegnern angenommen und gejagt werden.

Um solchen unangenehmen Situationen – oft mit Folgen – aus dem Weg zu gehen, verzichtete mancher Jagdflieger auf seine »Visitenkarte« am Ruder, während andere wiederum, sei es aus Starrköpfigkeit, Stolz, Selbstüberzeugung oder als Vorbild für jüngere Piloten von dieser Praxis bis Kriegsende nicht abgehen. Dieses Zugeständnis an die Überlegenheit des Gegners bleibt solchen Piloten erspart, die erst gar nicht mit dem Markieren ihrer Abschüsse begonnen hatten, darunter solche bekannte Leute wie Batz, Bühligen, Rall, um nur einige zu nennen.

Es sei hier eingefügt, daß das Buchen von abgeschossenen Gegnern während des 2. Weltkrieges nicht nur bei der Luftwaffe üblich war. Vielmehr sind bei allen kriegsführenden Nationen solche individuellen Markierungen zu finden. Im Gegensatz zur deutschen Methode, wird jedoch als Platz zur Anbringung der Abschußsymbole bei den Alliierten nicht das Seitenleitwerk, sondern die Rumpfseite unterhalb der Kabine gewählt. In Anlehnung an die Markierungspraxis bei der verbündeten Deutschen Luftwaffe erscheinen die Abschußkonten bei Italienern, Ungarn, Rumänen, Kroaten, Bulgaren, Finnen usw. ebenfalls auf den Seitenrudern. Die beistehenden Fotos mögen einen kleinen Überblick über diese Einzelheiten bei anderen Luftmächten geben.

Bereits sehr früh im 2. Weltkrieg gehen die Piloten der Luftwaffe dazu über, die Art der einzelnen Erfolge durch besondere Symbole voneinander zu unterscheiden. So werden Bodenzerstörungen von gegnerischen Flugzeugen durch nach unten spitz zulaufende Balken (s. Balthasar, Victor Bauer), oder gar auf der Fosse des Höhenruders (s. Hachfeld) angedeutet. Abgeschossene Sperr- oder Beobachterballone erscheinen in Form einer Ballon-Sil-

Lt. Col. John C. Meyer von der 8th U.S. Air Force in seiner North American P-51D »Mustang«. Die Hakenkreuze unter der Kabine zeigen 24 Luftsiege und $13^1/_2$ Bodenzerstörungen an. 6. Jan. 1945, Süd-England.

Lt. Col. John C. Meyer of the 8th Air Corce, USAAF, in the cockpit of his North American P-51D »Mustang«. The swastikas on its flanks indicate 24 aerial victories and $13^1/_2$ aircraft destroyed on the ground. 6 January 1945, Southern England.

Unter den erfolgreichsten amerikanischen Jagdfliegern steht Capt. David McCampbell an 4. Stelle. Die Siegeszeichen am Rumpf seiner Grumman F6F-5 »Hellcat« zeigen, daß er seine Erfolge im Pazifik gegen japanische Gegner errang.

Among the most successful American fighter aces Capt. David McCampbell holds 4th place. The victory marks on the fuselage of his Grumman F6F-5 »Hellcat« show that his successes were achieved against Japanese aircraft in the Pacific.

ships. In case of the latter the location of the placed hit was marked by shading of a different color (as with Liesendahl) and in most cases the size of the ship in tons was also marked.

Understandably enough fighter pilots were particularly proud of victories against four-engined bombers they succeeded shooting out of their formations and displayed the destruction of such a »heavy« in a special way.

In most cases a plan view of the bomber was either painted inside the »kill bar« (as with Schroer), or the aircraft silhouette itself marking the »kill« (as with Schnaufer) or the »kill bar« itself being devided vertically (as with Wurmheller) without, however, counting double.

The increase of night missions by the RAF Bomber Command over German territory and the resulting formation of night fighting units also had an influence upon the rudder markings. As with day victories no fixed rules existed and there appeared a multitude of markings signifying night victories. The most simple way of differentiating between day and night victories lay in marking the day victories with white bars and night victories with black (as with Kociok). The majority of night fighter pilots, however, marked a night victory by one or more slanted lines in a white bar (as with Ehle, Fellerer, Huelshoff, etc.). Since night fighter pilots had as enemies primarily two- or four-engined bombers it became a small sensation to catch a »Mosquito« night fighter since it was markedly superior in speed to the own aircraft and such a victory was expecially marked (as with F. Mueller's 23rd »kill«), similar to the day fighters and their four-engined »heavies«.

German aircraft of this period, crashing or forced down over enemy territory were favorites of souvenir hunters and the markings on the fabric-covered rudders were particularly valuable. Quite often these parts disappeared completely prior to the officials' arrival who had to be content with the rest. Considerably more difficult were things for German souvenir specialists who had to cut large pieces of metal from British or American aircraft in order to obtain the markings. In addition, they risked their hides since such activities

houette, zerstörte Panzerfahrzeuge als Panzer-Schattenriß, versenkte oder getroffene Schiffe als Schiffsumriß auf den Rudern. Dabei wird bei letzteren die Trefferlage auf dem angegriffenen Schiff (s. Liesendahl) mitunter durch besondere Schattierungen festgehalten, meist dazu auch die Bruttoregistertonnen des Schiffes.

Verständlicherweise sind die Jagdflieger auf ihre Abschußerfolge aus Viermot-Verbänden besonders stolz, und stellen die Vernichtung eines »Möbelwagens« vielfach gegenüber ihren anderen Abschüssen besonders heraus. Meist wird dann die Draufsicht-Silhouette eines Viermot in den Abschußbalken gemalt (s. Schroer), die Flugzeugsilhouette selbst steht für einen Abschuß (s. Schnaufer), oder der Abschußbalken wird vertikal gespalten (s.Wurmheller), ohne daß damit ein Viermot als zwei Abschüsse zählte!

Mit dem Anwachsen der Nachtangriffe des britischen Bomber Command auf deutsches Reichsgebiet, und die daraus resultierende Aufstellung von Nachtjagdverbänden, bildet sich die Besonderheit des Markierens von NJ-Abschüssen heraus. Wie bei Tagabschüssen besteht auch hierbei keine starre Regel, und es gibt eine Unzahl von Varianten zur Kennzeichnung von NJ-Erfolgen. Die einfachste Form der Unterscheidung zwischen Tag- und Nachtabschüssen besteht darin, Tag-Abschüsse durch weiße, solche bei Nacht durch schwarze Balken (s. Kociok) zu kennzeichnen. Der Großteil der Nachtjäger zeigt jedoch einen NJ-Abschuß durch einen oder mehrere schwarze Schrägstriche im weißen Balken an (s. Ehle, Fellerer, Hülshoff etc.). Da es die Nachtjäger bei ihren Gegnern in erster Linie mit zwei- oder viermotorigen Bombern zu tun haben, ist es bei diesen Verbänden eine kleine Sensation einen gegnerischen Nachtjäger, eine »Mosquito« zu erwischen, zumal es sich bei diesem Flugzeugmuster um einen an Schnelligkeit dem Nachtjäger weit überlegenen Kontrahenten handelt. Eine »Mosquito« wird deshalb bei den Nachtjägern (s. F. Müller, 23er Abschuß) mit dem gleichen Stolz wie bei den Tagjägern ein Viermot besonders markiert.

Ab- oder angeschossenen deutschen Maschinen, die in diesen Jahren auf gegnerischem Territorium niedergehen, gilt das besondere Augenmerk der Andenkenjäger. Dabei scheinen die Markierungen auf den stoffbespannten Seitenrudern der Jagdflugzeuge eine besondere Anziehungskraft auf die Souvenir-Hunters auszuüben; sehr oft verschwinden solche Flugzeugteile spurlos, bevor ein Bergungskommando an Ort und Stelle eintrifft und sich mit den verbliebenen Resten zufrieden geben muß.

Wesentlich schwieriger haben es dagegen die deutschen »Spezialisten« dieser Sparte, die sich mit dem

Finnlands erfolgreichster Jagdflieger Eino Ilmari Juutilainen mit 16 Abschußbalken an seiner Bf 109 G.

Finland's most successful fighter pilot Eino Ilmari Juutilainen with 16 »kill« bars on his Bf 109 G's stabilizer.

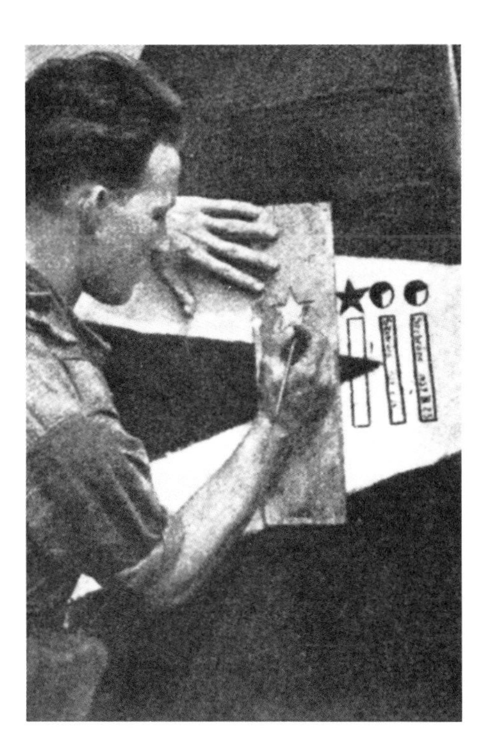

Auch die ungarischen Jagdflieger erledigten die Buchhaltung auf den Seitenrudern ihrer Jagdflugzeuge.

The Hungarian fighter pilots too kept book on their aircraft's vertical rudders.

were strictly forbidden. A number of rudders of German fighter aces coming down over friendly territory were saved from the wreckage and kept as mementos in officers' clubs or operations rooms. In this way a small number of rudders have survived the war and are today kept either in museums, by former pilots or their relatives.

In addition to marking victories on their aircraft many pilots of World War 2 continued the custom of WW 1 pilots in keeping a »victory« stick« in which each victory was marked by a notch or a ring.

Unit scores were often painted on the unit's operations trailer (as with the 9./JG 54) or marked on an elaborate scoreboard in the operations room. The pilot achieving the unit's 100th, 500th or 1000th victory was receiving an elaborate »red carpet« treatment upon his return, including an oversized garland of flowers or a plywood shield marking the occasion and had to endure a large and rather »wet« welcoming ceremony. Often the lucky pilot received a special certificate by the unit. Some of these certificates are still in existence and are guarded by their owners like their own eyes.

Similar to the fighter pilots other parts of the German forces too were keeping score of their destroyed aircraft and tanks. By its nature the antiaircraft batteries were next to the fighter units getting the lion's share of enemy aircraft »kills«. In the beginning each aircraft shot down merited a ring around the gun barrel but when, at a later date, armor shields were introduced the markings were applied to these. With the ships of the Kriegsmarine, the German navy, »kill« scores were often kept on the bridge or, in case of submarines, on the sides of the conning tower. Here, as well as in case of the flying units, personnel and material losses were climbing during the last war years to such an extent that it became impossible, simply for time reasons, to apply these markings to the new equipment, replacing the used up or replaced. Towards the end of the war there were hardly any »kill bars« kept on the aircraft of the fighter units that were flown by different pilots each day in a desparate last-ditch effort.

With the costly operation »Bodenplatte«

Ablösen oder Herausschneiden solcher Sondermarkierungen aus den Ganzmetallrümpfen englischer und amerikanischer Flugzeuge herumplagen müssen, und dazu beim Erwischtwerden noch Kopf und Kragen riskieren.

Eine Anzahl von Rudern deutscher Jagdflieger-Asse, die über eigenem Gebiet abstürzen oder notlanden müssen, werden von den Bergungskommandos der betreffenden Einheiten aus dem Bruch ausgebaut und in Kasinos oder Gefechtsständen aufbewahrt. Auf diese Art haben einige, wenige Ruder den Krieg überlebt und befinden sich heute in Museen, im Besitz damaliger Piloten oder deren Angehörigen.

Neben dem Aufmalen von Abschüssen auf die eigene Maschine übernehmen viele Flugzeugführer des 2. Weltkrieges den Brauch ihrer Fliegerkameraden von 1914/18 sich einen Abschußstock zuzulegen, auf dem für jeden errungenen Luftsieg eine Kerbe oder Ring eingeschnitten wird.

Die Abschüsse einer ganzen Einheit werden oft auf den Gefechtsanhänger aufgepinselt (s. 9./JG 54), oder auf einer künstlerisch gestalteten Abschußtafel im Gefechtsstand festgehalten. Der Flugzeugführer, der einen Jubiläumabschuß, den 100ten., 500ten. oder 1000ten. eines Verbandes erzielt, wird bei der Landung in der Regel mit großem Halloh, einem überdimensionalen Laubkranz oder Sperrholzschild empfangen, und muß eine gewaltige Gratulationszeremonie, bei der unter sichtlichem Wohlbehagen aller Beteiligten Mengen belebender Getränke verkonsumiert werden, über sich ergehen lassen. Oft wird dem Jubiläumsschützen ein besonderes Diplom zur Erinnerung vom Verband ausgestellt. Auch einige dieser Diplome existieren heute noch und werden von ihren Besitzern wie ein Augapfel gehütet.

Ähnlich wie bei den Jagdfliegern werden auch bei anderen Wehrmachtsteilen Flugzeug-, Panzer- oder sonstige Abschüsse gebucht. Naturgemäß ist die Flak neben den Jagdverbänden die Waffe, die an den Abschüssen gegnerischer Flugzeuge am stärksten beteiligt ist. Anfangs wird von den Flak-Kanonieren für jeden abgeschossenen Gegner ein Ring um das Geschützrohr gemalt; bei Um- oder Nachrüstung der Fliegerabwehrkanonen mit Panzerschutzschilden werden jedoch späterhin die Abschußerfolge auf diese Schutzschilder markiert. Bei der Kriegsmarine finden sich die Abschußbuchungen vielfach auf den Verkleidungsblechen der Kommandobrücke, bei den Unterseebooten auf den Turmflanken.

Hier, wie bei den fliegenden Verbänden steigen neben personellen die Materialverluste in den letzten Kriegsjahren so sehr an, daß es aus zeitlichen Gründen einfach unmöglich wird, diese Sondermarkie-

Dieser italienische Pilot nimmt es genau mit der Identifizierung seiner Abschüsse, wie die exakte Silhouette einer »Spitfire« beweist.

This Italian pilot is paying close attention as to the identity of his victims as the exact outline of a »Spitfire« proves.

Abschußmarkierungen auf der Brücke eines Minenlegers der »Kanalflotille« der Kriegsmarine. Die dunklen Silhouetten zeigen bestätigte, die hellen unbestätigte Abschüsse an. Sept. 1941, Cherbourg.

Victory marks on the bridge of a mine layer of the Kriegsmarine's »Kanalflotille« (Channel Flotilla). The dark silhouettes are confirmed »kills«, the bright are unconfirmed ones. September 1941, Cherbourg, France.

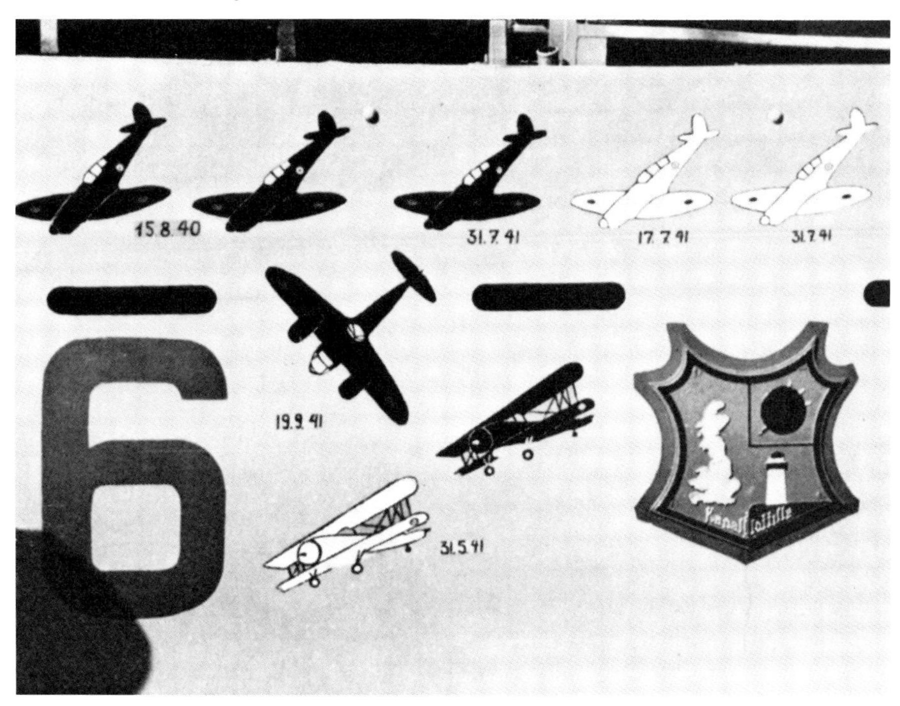

(Ground Plate) early in 1945 the Luftwaffe's final hours are drawing nearer, a fact that even some astonishing aerial victories could not prevent. The last chapter of the air war was written by the »storm groups«, the jet and rocket fighter units of the home defence, that are mounting a final, almost suicidal effort to stem the tide of the Allied bomber and fighter streams, but without having the slightest chance at all.

rungen auf dem schnell verbrauchten und durch neue Serien ersetzten Gerät anzubringen. So finden sich gegen Kriegsende kaum noch Abschußmarkierungen auf den Flugzeugen der Jagdgruppen, in denen die verbliebenen Maschinen täglich von anderen Piloten im aufreibenden Einsatz geflogen werden.

Mit Beginn des Jahres 1945 kündigt sich mit der verlustreichen Aktion »Bodenplatte« die endgültige Niederlage der Luftwaffe an, ein Faktum, über das auch die bisweilen noch erstaunlichen Abschußerfolge nicht hinwegtäuschen können. Das letzte Kapitel des Luftkrieges wird von den »Sturmgruppen«, Turbo- und Raketenjägereinheiten der Reichsverteidigung geschrieben, die in nahezu selbstmörderischem Einsatz gegen die Ströme von schweren, mittleren und Jagdbombern anrennen, ohne jedoch die geringste Chance zu haben das Blatt noch einmal wenden zu können.

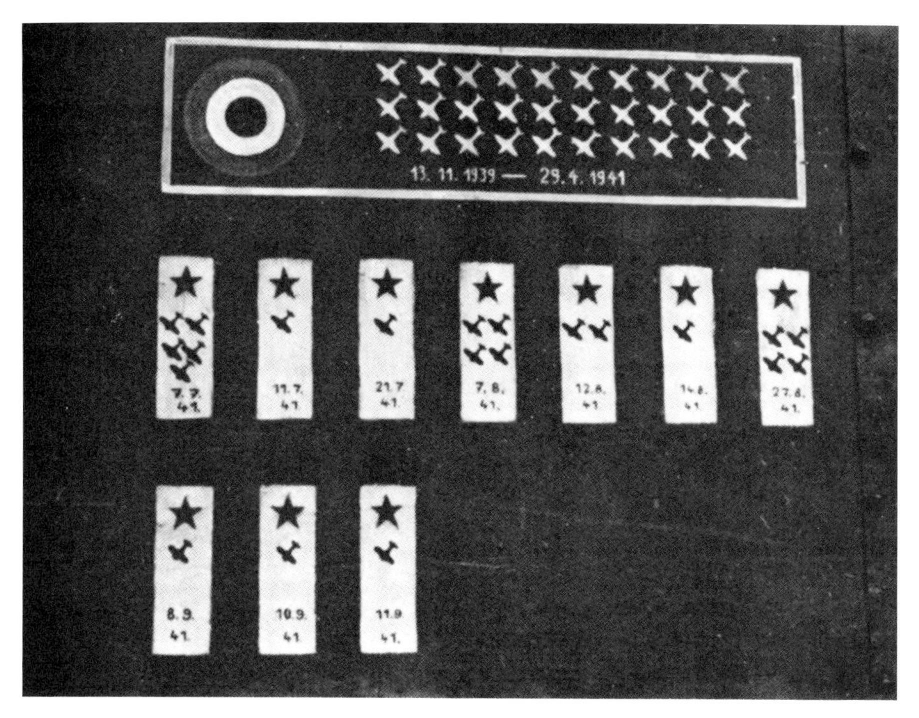

Die stattliche Abschußliste einer Flakbatterie auf dem Gefechtsanhänger der Einheit. Ostfront, September 1941.

The impressive »kill« score of an antiaircraft battery on one of the unit's trailers. Eastern front, September 1941.

Panzerschutzschild einer 8,8 cm Flak mit Erfolgssymbolen für vernichtete Flugzeuge, Panzer, Bunker, Geschütze und Beobachtungsstände.

Armor shield of an 8.8 cm antiaircraft gun with victory symbols for destroyed aircraft, tanks, shelters, guns, and observation points.

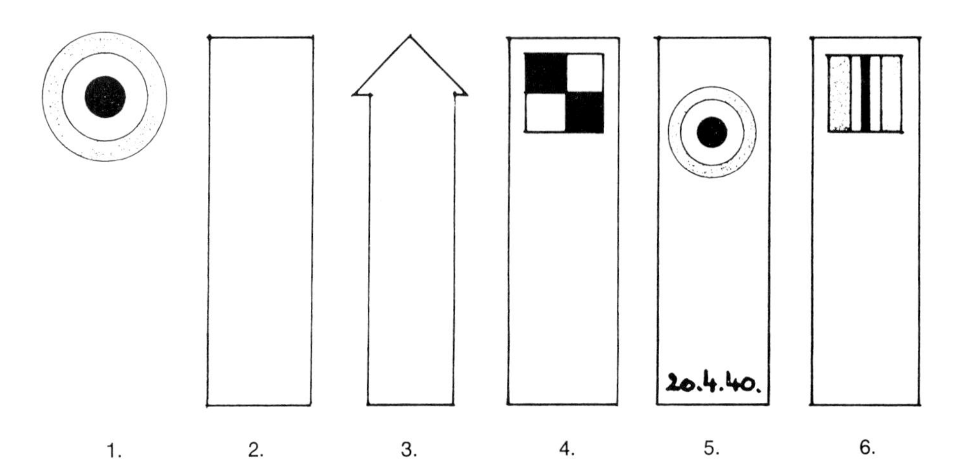

1. Luftsieg, über einen französ. Gegner (Nur Kokarde)
2. Luftsieg, ohne nähere Angabe
3. Luftsieg, zur Unterscheidung gegenüber Bodenzerstörung
4. Luftsieg, über poln. Gegner
5. Luftsieg, über französ. Gegner, mit Datum
6. Luftsieg, über norwegischen Gegner

1. Aerial victory over a French adversary (roundel only).
2. Aerial victory without further explanation
3. Aerial victory differing from ground destruction
4. Aerial victory over Polish adversary.
5. Aerial victory over French adversary with date.
6. Aerial victory over Norwegian adversary.

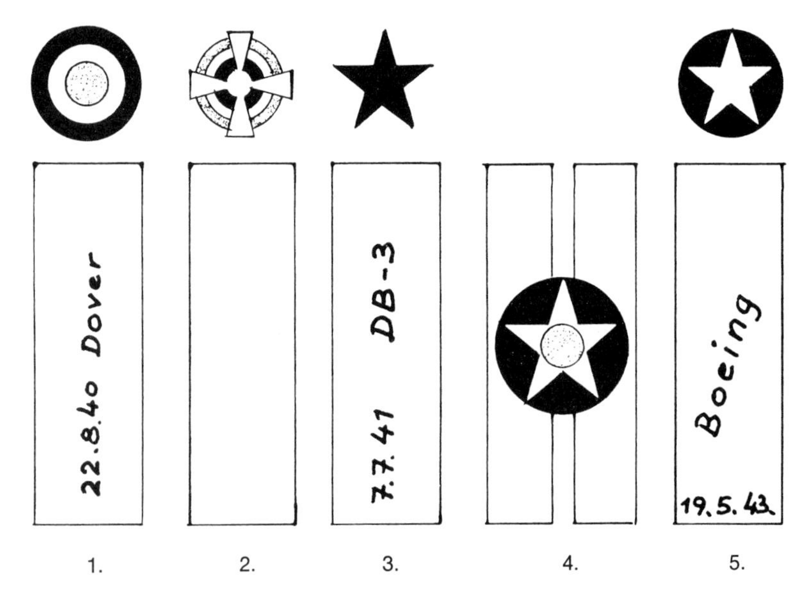

1. Luftsieg, über englischen Gegner, mit Datum und Abschußort
2. Luftsieg, über jugoslawischen Gegner
3. Luftsieg, über russischen Gegner mit Datum und Typ
4. Luftsieg, über U.S. Viermot
5. Luftsieg, über U.S. Gegner mit Datum und Typ

1. Aerial victory over British adversary with date and location
2. Aerial victory over Yugslav adversary
3. Aerial victory over Russian adversary with date and type
4. Aerial victory over four-engined US bomber
5. Aerial victory over US adversary with date and type

Zehn Luftsiege über russische Gegner zusammengefaßt.
Ten aerial victories over Russian adversaries combined.

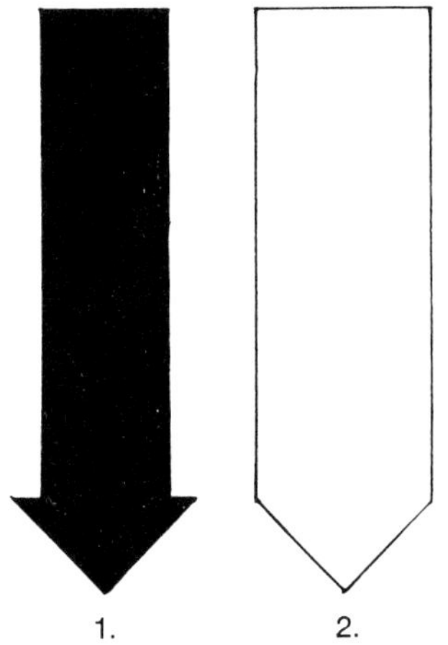

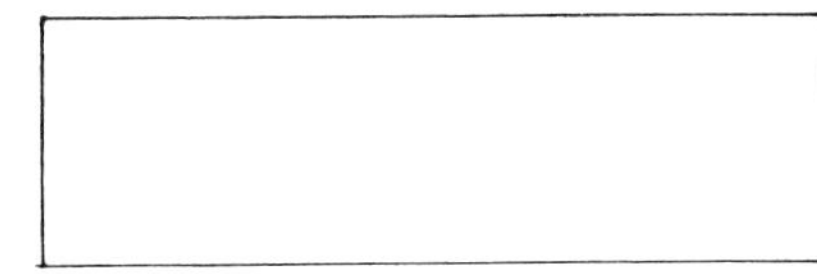

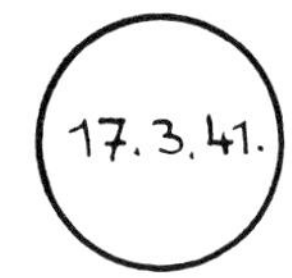

1. 2. 3.

1. Bodenzerstörung ohne nähere Angabe
2. Bodenzerstörung ohne nähere Angabe
3. Bodenzerstörung mit Datum

1. Ground destruction without further explanation
2. Ground destruction without further explanation
3. Ground destruction with date.

1. 2. 3. 4. 5.

1. Nacht-Luftsieg, ohne nähere Angabe
2. Nacht-Luftsieg, über englischen Gegner
3. Nacht-Luftsieg, über englischen Gegner
4. Nacht-Luftsieg, über englischen Mehrmot.
5. Nacht-Luftsieg, über englischen Viermot., mit Datum

1. Night victory without further explanation
2. Night victory over British adversary
3. Night victory over British adversary.
4. Night victory over multi-engined British bomber
5. Night victory over multi-engined British bomber with date

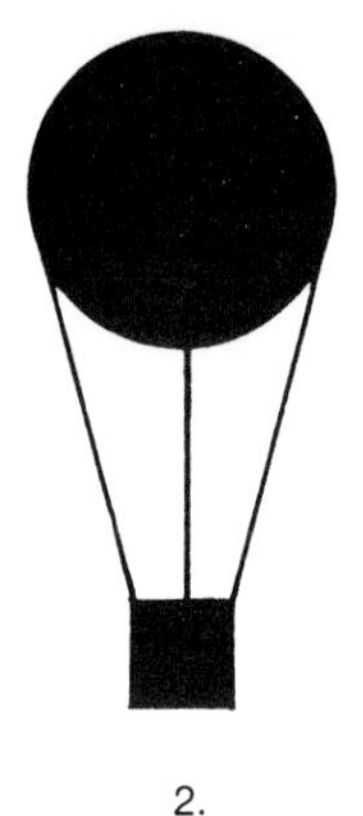

1. 2. 3.

1. Abschuß eines Sperrballones
2. Abschuß eines Beobachtungsballones
3. Abschuß eines gegnerischen Panzers

1. Destruction of a barrage balloon.
2. Destruction of an observation balloon
3. Destruction of an enemy tank

Versenkung eines gegnerischen 1000 t-U-Bootes.

Sinking of an enemy 1000 ton submarine.

Versenkung eines gegnerischen Frachters, mit Datum und Schiffs-Tonnage.

Sinking of an enemy freighter with date and ship size.

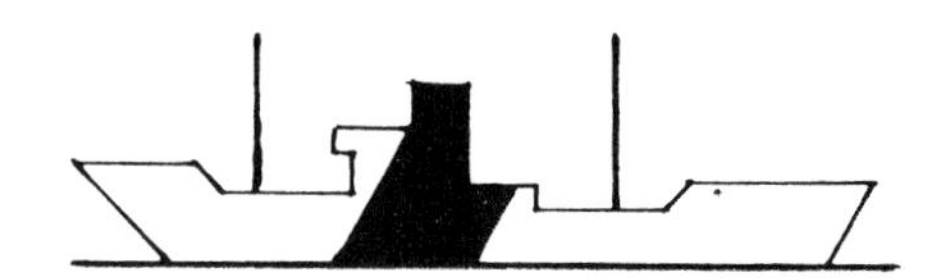

Versenkung eines gegnerischen Frachters durch Treffer mittschiffs.

Sinking of an enemy freighter by a hit midships.

Bär, Heinz 1./JG 51 Bf 109 E
Pihen/Frankr. 2. Sept. 1940
8. Luft-Sieg (Gesamt-Luftsiege 220)

Bär, Heinz 1./JG 51 Bf 109 E
Coquelle/Frankr. 21. Apr. 1941
15. L.S. (Ges. 220)

Bär, Heinz
Kertsch/Ostfront
103. L.S. (Ges. 220)

Kdr. IV./JG 51
19. Mai 1942

Bf 109 F

Bär, Heinz
Kdr. IV./JG 51
Bf 109 F
Kertsch IV/Ostfr.
27. Juni 1942
113.L.S. (Ges. 220)

Bär, Heinz
Staffelführ. II./JG 1
FW 190 A
Störmede
22. Apr. 1944
200 L.S. (Ges. 220)

Balthasar, Wilhelm Kdr. III./JG 3
Frankreich Anfang Sept. 1940
25. L.S. + 13 Bodenzerstörungen (Ges. 47)
links: Günther Lützow

Bartels, Heinrich
IV./JG 27
Bf 109 F
Nisch/Jugosl.
23. Okt. 1943
56 L.S. (99)

Bartels, Heinrich
IV./JG 27
Bf 109 F
Kalamaki/Griechenland
17. Nov. 1943
70. L.S. (99)

Bartels, Heinrich
IV./JG 27
Bf 109 G
Götzendorf b/Wien
24. Apr. 1944
76. L.S. (99)

Bauer, Victor
III./JG 3
Bf 109 F
Tschugujew/Ostfr.
19. Mai 1942
51.L.S. (106)

Bauer, Victor	III./JG 3	Bf 109 F
Frolow vor Stalingrad	22. Juli 1942	
91. L.S. + 9 Bodenzerstör.	(106)	

Beißwenger, Hans
II./JG 54
Bf 109 F
Torrossowo/Ostfr.
23. Sept. 1941
30.L.S. (152)

Bertram, Otto
Kdr. III./JG 2
Bf 109 E
bei Cherbourg/Frankr.
9. Okt. 1940
13. L.S. (21)

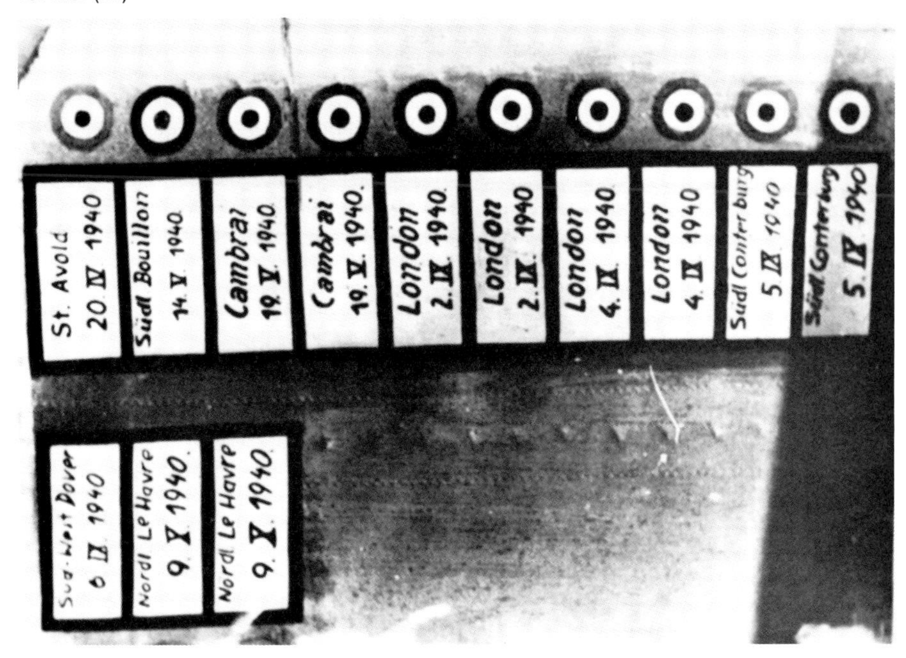

Bob, Hans-Ekkehard
Guines/Frankr.
14. L.S. (59)

Staffelkap. 9./JG 54
30. Sept. 1940

Bf 109 E

Bob, Hans-Ekkehard
St. Kap. 9./JG 54
Bf 109 E
Parndorf
März 1941
19. L.S. (59)

Börngen, Ernst | 5./JG 27 | Bf 109 F
60–80 km SO El Daba nach Notlandung | | 11. Juli 1942
8. L.S. (45)

Bohn, Paul | St. Kap. 4./NJG 2 | Ju 88 C
bei Mailand | 7. Juli 1941
5. L.S. (5)

Bonin, Eckart-Wilhelm von Kdr. II./NJG 1 Bf 110 G
St. Trond Febr. 1944
31. L.S. (39)

Boremski, Eberhard von
St. Kap. 7./JG 3
Bf 109 F
Frolow vor Stalingrad
22. Juli 1942
67. L.S. (90)

Brandt, Walter
I./JG 77
Bf 109 F
Comiso/Italien
Aug. 1942
30. L.S. (57)

Brändle, Kurt
Kdr. II./JG 3
Bf 109 F
Frolow vor Stalingrad
Juli 1942
95. L.S. (180)

Brandis, Felix (rechts)	St. Kap. 10.(Z)/JG 5	Bf 110
Rovaniemi/Finnland	Winter 1941/42	
18. L.S. (Gesamt ?)	links: Bordf. Baus	

Broennle, Herbert	4./JG 54	Bf 109 F
Siwerskaja/Ostfr.	16. März 1942	
24. L.S. (57)		

Busch, . . . | I./JG 51 | Bf 109 F
Ponjatowka/Ostfr. | ca. 16. Okt. 1941 |
20. L.S. (?)

Busch, . . .	I./JG 51	Bf 109 F
Ponjatowka/Ostfr.	ca. 16. Okt. 1941	
20. L.S. (?)		

Clerico, Max	7./JG 54	Bf 109 E
Guines/Frankr.	Herbst 1940	
7. L.S. (?)		

Dahl, Walther
Adj. JG 3
Bf 109 F
Ostfront
Herbst 1942
50. L.S. (128)

Dahl, Walther
III./JG 3
Bf 109 G
Bad Wörishofen
Winter 1943
55. L.S. (128)

Dahmer, Hugo
Luostari/Finnl.
22. L.S. (57)

1./JG 77
11. Juli 1941

Bf 109 E

Dammers, Hans
7./JG 52
Bf 109 F
Bjelgorod/Ostfr.
Mai 1943
100. L.S. (113)

Demuth, Erich
I./JG 1
He 162 A
Leck/Holstein
Mai 1945
16. L.S. (16)

Dinger, Fritz
St. Kap. 4./JG 53
Bf 109 F
Tunesien
Frühjahr 1943
53. L.S. (67)

Döbrich, Hans Petsamo/Finnl. 26. L.S. (65)	6./JG 5 Okt. 1942	Bf 109 F

Drewes, Martin (links) Leeuwarden/Holland 47. L.S. (52)	Kdr. III./NJG 1 21. Juli 1944	Bf 110G

Düllberg, Ernst
Kdr. III./JG 27
Bf 109 G
Wiesb.-Erbenheim
März 1944
27. L.S. (50)

Düllberg, Ernst (links)
Wiesb.-Erbenheim
27. L.S. (50)
Kdr. III./JG 27
März 1944
rechts: Werner Schroer
Bf 109G

Eckerle, Franz
Kdr. I./JG 54
Bf 109 F
Siwerskaja/Ostfr.
Sept. 1941
30. L.S. (59)

Ehle, Walter (links)
Stade
9. L.S. (36)
Kdr. II./NJG 1
30. Juni 1941
rechts: Bordf. Weng
Bf 110E

Ehrler, Heinrich
Petsamo/Finnl.
32. L.S. (205)

6./JG 5
Juli 1942

Bf 109 F

Ehrler, Heinrich Petsamo/Finnl. 77. L.S. (205)	St. Kap. 6./JG 5 27. März 1943	Bf 109 G

Falck, Wolfgang Aalborg/Dänemark 8. L.S. (8)	Kdr. I./ZG 1 9. Apr. 1940	Bf 110 C

Fellerer, Leopold (Mitte) Kdr. II./NJG 5 Bf 110G
Parchim 14. Jan. 1944
22. L.S. (41) links: Bordf. Hätscher

Fellerer, Leopold
Kdr. II./NJG 5
Bf 110 G
Parchim
14. Jan. 1944
22. L.S. (41)

Fleig, Erwin Star-Bychow/Ostfr. 25. L.S. (66)	2./JG 51 20. Juli 1941	Bf 109 F

Galland, Adolf Audembert/Frankr. 56. L.S. (104)	Kdore. JG 26 28. Nov. 1940	Bf 109 E

Galland, Adolf
Audembert/Frankr.
57. L.S. (104)

Kdore. JG 26
5. Dez. 1940

Bf 109 E

Galland, Adolf
Kdore. JG 26
Bf 109 F
Audembert/Frankr.
aufgen.: 5. Dez. 1941
94. L.S. (104)

Gildner, Paul
Leeuwarden/Holland
33. L.S. (48)

II./NJG 2
9. Juni 1942

Do 17Z-10

Glunz, Adolf
II./JG 26
FW 190 A
Beauvais/Frankr.
31. Aug. 1943
39. L.S. (71)

Glunz, Adolf
II./JG 26
FW 190 A
Cambrai-Epinoy/Frankr.
31. Dez. 1943
54. L.S. (71)

Glunz, Adolf
II./JG 26
Fw 190 A
Cambrai-Epinoy/Frankr.
22. Febr. 1944
58. L.S. (71)

Gollob, Gordon	Kdr. II./JG 3	Bf 109 F
Stschastliwaja/Ostfr.	21. Aug. 1941	
33. L.S. (150)		

Gollob, Gordon	Kdore. JG 77	Bf 109 F
Oktoberfeld/Krim	21. Juni 1942	
107. L.S. (150)		

Graf, Hermann St. Kap. 9./JG 52 Bf 109 G
Krim/Ostfr. 2. Okt. 1942
202. L.S. (212)

Graf, Hermann
St. Kap. 9./JG 52
Bf 109 G
Krim/Ostfr.
2. Okt. 1942
202. L.S. (212)

Graf, Hermann
Wiesb.-Erbenheim
205. L.S. (212)

Kdr. J. Gr. 50
6. Sept. 1943

Bf 109 G

Gratz, Karl
8./JG 52
Bf 109 G
Krim/Ostfr.
Herbst 1942
83. L.S. (138)

Grislawski, Alfred
J. Gr. 50
Bf 109 G
Wiesb.-Erbenheim
6. Sept. 1943
112. L.S. (133)

Hachfeld, Wilhelm
3./LG 2
Bf 109 E
Mt. Ecourez/Frankr.
29. Mai 1940
2. L.S. (11)
6. Bodenzerstör. (32)

Hackl, Anton
Rotenburg b/Bremen
141. L.S. (192)

Kdr. III./JG 11
März 1944

Fw 190A

Hafner, Anton
Orel-Nord/Ostfr.
62. L.S. (204)

6./JG 51
26. Aug. 1942

Bf 109 F

Hahn, Hans »Assi«
Kdr. III./JG 2
Bf 109 F
St. Pol/Frankr.
Juli 1941
31. L.S. (108)

Hahn, Hans von »Vadder« Kdr. I./JG 3 Bf 109 F
Berditschew/Ostfr. Juli 1941
24. L.S. (34) + 3 Bodenzerstör. + 3 Ballone

Hanke, Heinz (rechts) | 9./JG 1 | Fw 190A
Husum | 26. Febr. 1943
1. L.S. (9)

Hartmann, Ludwig | 9./JG 2 | Fw 190 A
Theville/Frankr. | Mai 1942
10. L.S. (Ges. ?)

Hartmann, Erich | 9./JG 52 | Bf 109 G
Nowo Saporoshje/Ostfr. | 2. Okt. 1943
121. L.S. (352)

Held, Alfred | II./JG 77 | Bf 109 E
Wangerooge | 4. Sept. 1939
1. L.S. (Ges. ?)

Hermann, Kurt
I./NJG 2
Ju 88 C
Gilze-Rijen
1941
9. L.S. (9)

Heyer, Hans-Joach.
Siwerskaja/Ostfr.
20. L.S. (53)

III./JG 54
Juni 1942

Bf 109 F

Hohagen, Erich
Chatalowka/Ostfr.
25. L.S. (55)

II./JG 51
11. Aug. 1941

Bf 109 F

Hrabak, Dietrich (rechts)
Kdr. II./JG 54
Bf 109 F
Mal Owsischtschi/Ostfr.
30. Juli 1941
24. L.S. (125)
links: Hans Philipp

Hübl, Rudolf
Schiphol/Holland
3. L.S. (16)

1./JG 1
Febr. 1943

Fw 190 A

Hübner, Werner
4./JG 51
Bf 109 E
Mardyck/Holland
16. Febr. 1941
7. L.S. (7)

Hülshoff, Karl Gilze-Rijen/Holland 4. L.S. (11)	Kdr. I./NJG 2 8. Apr. 1941	Ju 88 C

Huy, Wolf-Dietr. Jassy/Rumänien 7. L.S. (40) + 8 Schiffe	III./JG 77 Juli 1941	Bf 109 F

Huy, Wolf-Dietr.
III./JG 77
Bf 109 F
Jassy/Rumänien
Aug. 1941
14. L.S. (40)
+ 8 Schiffe

Ihlefeld, Herbert
Calais-Marc
32. L.S. (130)

Kdr. I./LG 2
13. März 1941

Bf 109 F

Ihlefeld, Herbert	Kdr. I./LG 2	Bf 109 F
Jassy/Rumänien	12. Juli 1941	
47. L.S. (130)		

Johnen, Wilhelm (Mitte)	5./NJG 5	Bf 110G
Parchim	Jan. 1944	
14. L.S. (34)	Links: Bordf. Kilian	rechts: Bordsch. Mahle

Jung, Harald | 4./JG 20 | Bf 109 E
Bönninghardt | 22. März 1940
1. L.S. (20)

Kempf, Karl | III./JG 54 | Bf 109 F
Siwerskaja/Ostfr. | 19. Sept. 1941
24. L.S. (65)

Kempf, Karl Siwerskaja/Ostfr. 50. L.S. (65)	III./JG 54 12. Juni 1942	Bf 109 F
Klein, Hans (Mitte) Tschaplinka/Ostfr. 9. L.S. (10)	9./JG 52 17. Okt. 1941	Bf 109 F

Koch, Harry 9./JG1 Bf 109 F
Bergen am Zee/Holland Jan. 1942
13. L.S. (Ges. ?) + 1 Schiff + 4 Jaboeinsätze

Kociok, Josef 10. (NJ)/ZG 1 Bf 110 F
Krim/Ostfr. Frühjahr 1943
20. L.S. (33)

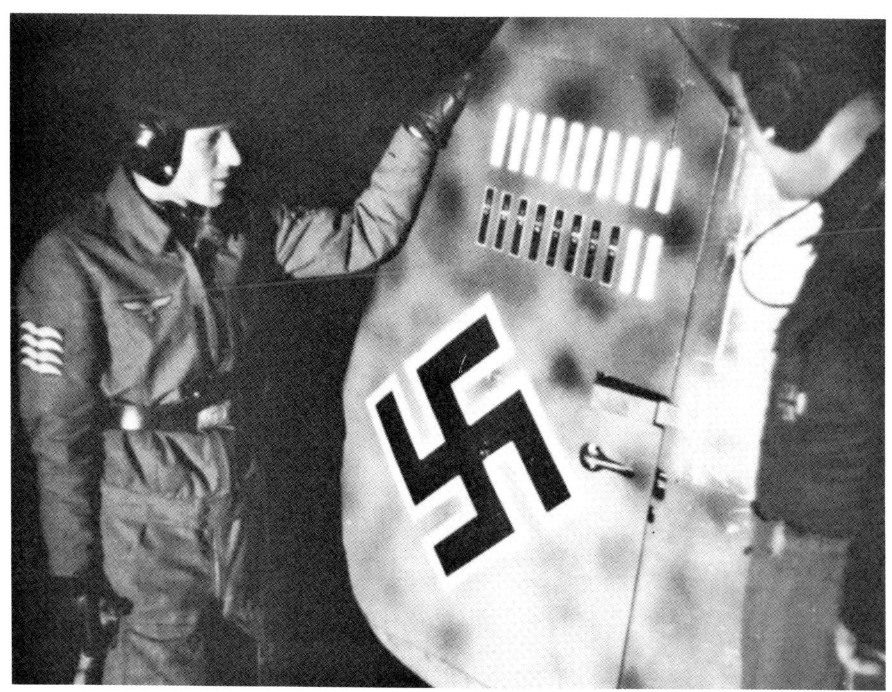

König, Hans-Heinrich | Kdr. I./JG 11 | Fw 190A
Husum | Apr. 1944
17. L.S. (28)

Kraft, Josef | III./NJG 6 | Bf 110 G
Steinamanger/Ungarn | 22. Aug. 1944
38. L.S. (56) | links: Bordsch. Pabst

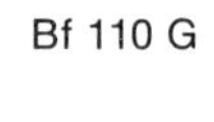

Kraft, Josef | III./NJG 6 | Bf 110 G
Steinamanger/Ungarn | 20. Okt. 1944
49. L.S. (56)

Kraft, Josef (Mitte)	III./NJG 6	Bf 110 G
Steinamanger/Ungarn	20. Okt. 1944	
49. L.S. (56)	links: Bordsch. Pabst	rechts: Bordf. Teubner

Krupinski, Walter
St. Kap. 7./JG 52
Bf 109 G
Stalino/Ostfr.
Aug./Sept. 1943
120. L.S. (197)

Lemke, Siegfried Diabolo/Italien 48. L.S. (96)	1./JG 2 17. März 1944	Fw 190A

Lent, Helmut Jever 4. L.S. (110)	1./ZG 76 18. Dez. 1939	Bf 110C

Lent, Helmut
II./NJG 2
Bf 110 F
Leeuwarden/Holland
26. Febr. 1942
30. L.S. (110)

Lent, Helmut
Kdr. IV./NJG 1
Bf 110 F
Leeuwarden/Holland
21. Jan. 1943
58. L.S. (110)

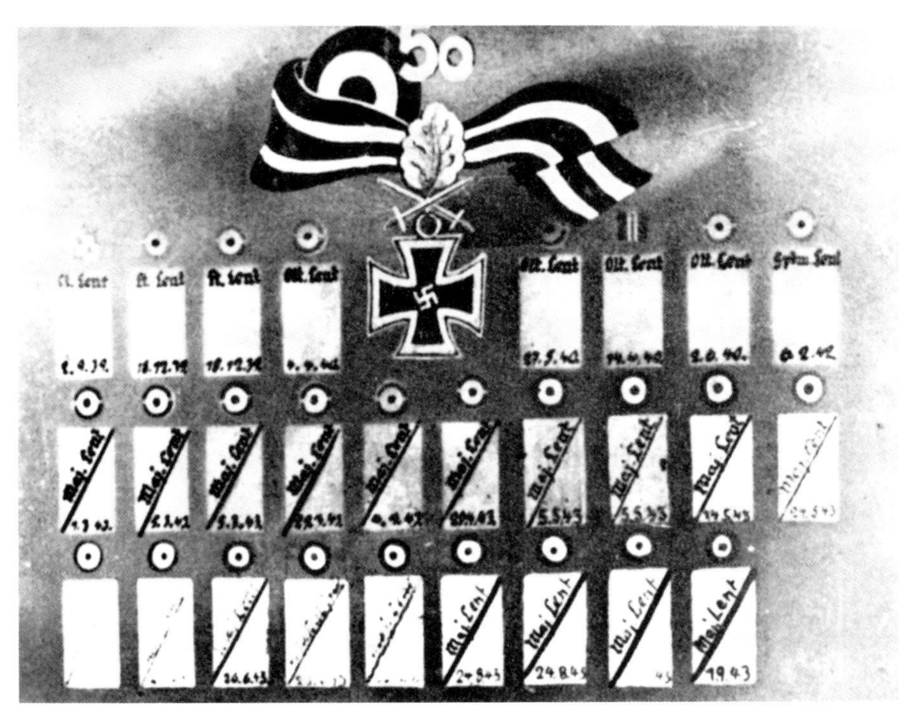

Lent, Helmut — Kdore. NJG 3 — Bf 110 G
Stade — 1. Sept. 1943
77. L.S. (110)

Litjens, Stefan (rechts) — 4./JG 53 — Bf 109 F
Lyuban b/Leningrad — 11. Sept. 1941
22. L.S. (38)

Lippert, Wolfgang | II./JG 27 | Bf 109 E
Balkan | 1941
21. L.S. (29)

Lützow, Günther | Kdore. JG 3 | Bf 109 F
Schatalowka/Ostfr. | 24. Okt. 1941
101. L.S. (108)

Machold, Werner St. Kap. 7./JG 2 Bf 109 E
Le Havre/Frankr. 30. Sept. 1940
26. L.S. (32)

Machold, Werner
St. Kap. 7./JG 2
Le Havre/Frankr.
30. Sept. 1940
26. L.S. (32)

Marseille, Hans-Joach.	3./JG 27	Bf 109 F
Gazala/Nordafrika	21. Febr. 1942	
50. L.S. (158)		

Marseille, Hans-Joach. St. Kap. 3./JG 27 Bf 109F
Gazala/Nordafr. 17. Juni 1942
101. L.S. (158)

Marseille, Hans-Joach. St. Kap. 3./JG 27 Bf 109F
Quotaifiya/Nordafr. 15. Sept. 1942
151. L.S. (158)

Marseille, Hans-Joach.
St. Kap. 3./JG 27
Bf 109 F
Quotaifiya/Nordafr.
30. Sept. 1942
158. L.S. (158)

Mayer, Egon
Kdr. III./JG 2
Fw 190 A
Beaumont/Frankr.
Juni 1943
68. L.S. (102)

Meckel, Helmut
Südabschnitt/Ostfr.
22. L.S. (mind. 25)

I./JG 3
Juli 1941

Bf 109 F

Mertens, Helmut
I./JG 3
Bf 109 F
Frelow vor Stalingrad
22. Juli 1942
47. L.S. (97)

Methfessel, Werner
Mannheim-Sandhofen
8. L.S. (Ges. ?)

14./LG 1
Mai 1940

Bf 110 C

Michalski, Gerhard
Kdr. II./JG 53
Bf 109 F
Comiso/Italien
Juli 1942
43. L.S. (73)

Mix, Erich — T.O. I./JG 53 — Bf 109 E
Darmstadt-Griesheim — Dez. 1939
2. L.S. (Ges. ?)

Mölders, Werner — Kdore. JG 51 — Bf 109 E
Pihen/Frankr. — 31. Aug. 1940
32. L.S. (115)

Mölders, Werner
Kdore. JG 51
Bf 109 F
Mardyck/Belgien
1. Dez. 1940
55. L.S. (115)

Mölders, Werner Kdore. JG 51 Bf 109 F
Düsseldorf Juni 1941
68. L.S. (115)

Mölders, Werner Kdore. JG 51 Bf 109 F
Terospol/Ostfr. 24. Juni 1941
73. L.S. (115)

Mölders, Werner
Kdore. JG 51
Bf 109 F
Star Bychow/Ostfr.
15. Juli 1941
101. L.S. (115)

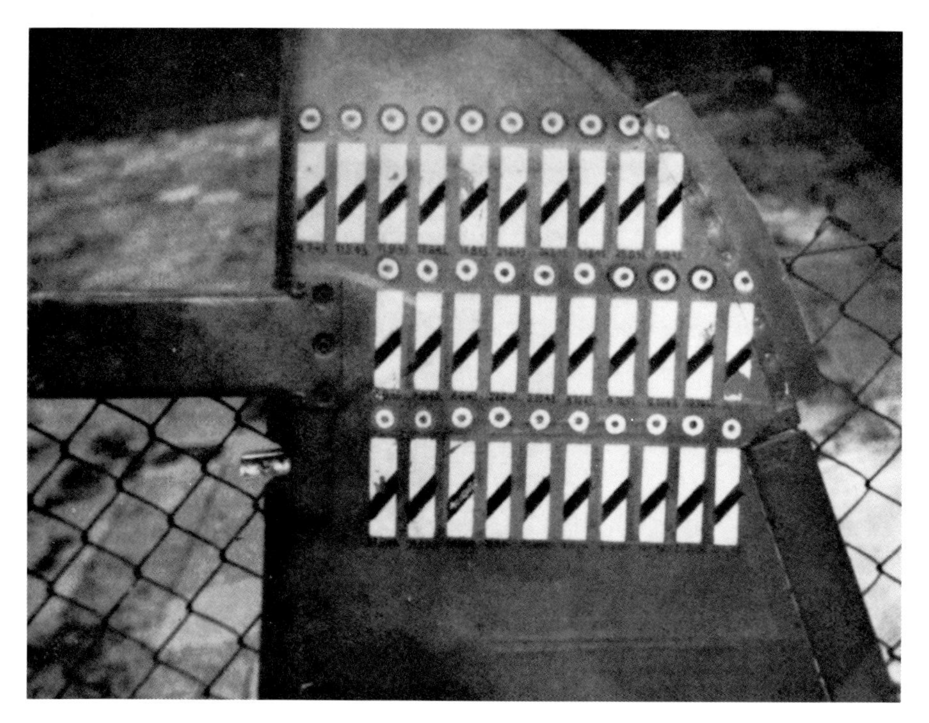

Müller, Friedr.-Karl Werneuchen 30. L.S. (30)	Kdr. I./NJG 11 13. Sept. 1944	Bf 109 G
Müncheberg, Joachim Caffiers/Frankr. 17. L.S. (135)	St. Kap. 7./JG 26 6. Sept. 1940	Bf 109E

Müncheberg, Joachim
St. Kap. 7./JG 26
Bf 109 E
Gela/Sizilien
28. März 1941
33. L.S. (135)

Müncheberg, Joachim
Kdr. III./JG 26
Fw 190 A
Coquelles/Frankr.
8. Dez. 1941
61. L.S. (135)

Mütherich, Hubert
St. Kap. 5./JG 54
Bf 109 F
Mal Owsischtschi/Ostfr.
Ende Aug. 1941
37. L.S. (43)

Mütherich, Hubert (links)
St. Kap. 5./JG 54
Bf 109 F
Mal Owsischtschi/Ostfr.
Ende Aug. 1941
37. L.S. (43)
rechts: Josef Pöhs

Nöcker, . . . (links) | 3./JG 1 | Bf 109 E
de Kooy/Holland | Frühjahr 1941 |
2. L.S. (Ges. ?)

Olejnik, Robert | St. Kap. 4./JG 1 | Bf 109 F
Düsseldorf | April 1942 |
37. L.S. (41)

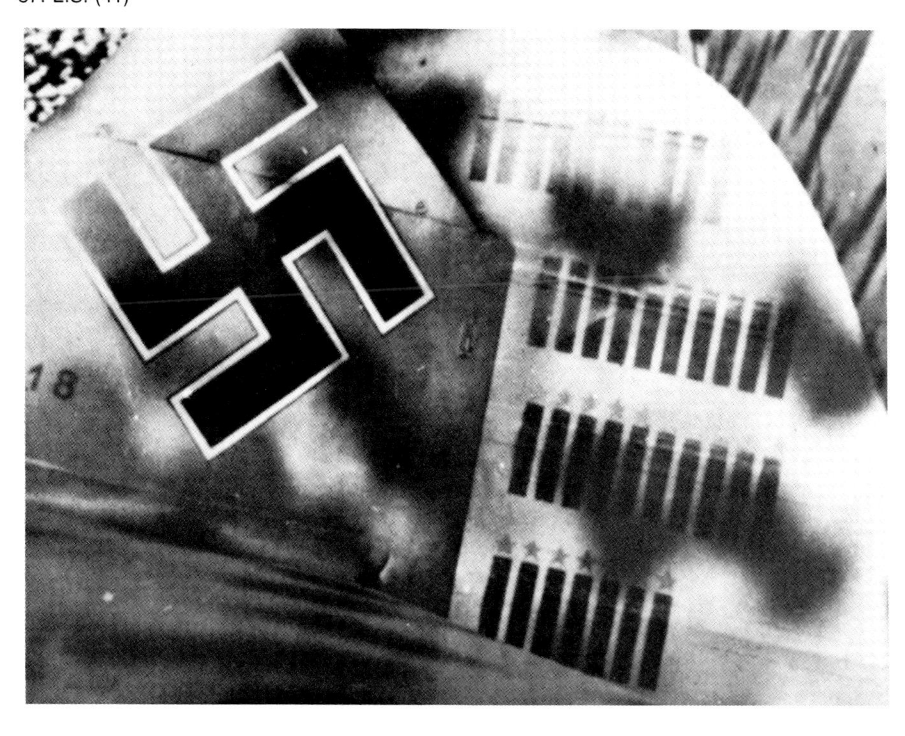

Ostermann, Max-Hellmuth
St. Kap. 7./JG 54
Bf 109 F
Siwerskaja/Ostfr.
12. Mai 1942
100. L.S. (102)

Philipp, Hans
St. Kap. 4./JG 54
Bf 109 F
Mal Owsischtschi/Ostfr.
Anfang Aug. 1941
57. L.S. (206)

Philipp, Hans — Kdr. I./JG 54 — Bf 109 F
Krasnogwardeisk/Ostfr. — März 1942
90. L.S. (206)

Philipp, Hans — Kdr. I./JG 54 — Bf 109 F
Krasnogwardeisk/Ostfr. — 31. März 1942
100. L.S. (206)

Philipp, Hans | Kdr. I./JG 54 | Bü 131
Krasnogwardeisk/Ostfr. | Febr. 1942
Reiseflugzeug | (Personal courier aircraft)

Pichler, Johann | 7./JG 77 | Bf 109 F
Simfereopol/Krim | 15. Nov. 1941
15. L.S. (52) + 2 Schiffe

Priller, Josef St. Kap. 6./JG 51 Bf 109 E
Mardyck/Belgien 17. Okt. 1940
20. L.S. (101) + 2 Bodenzerstör.

Priller, Josef Kdr. III./JG 26 Fw 190 A
Wevelghem/Belgien Apr./Mai 1942
ca. 70. L.S. (101)

Priller, Josef Kdr. III./JG 26 Fw 190 A
Wevelghem/Belg. 1. Juni 1942
73. L.S. (101)

Rohwer, Detlev I./JG 3 Bf 109 F
Bialacerkiew/Ostfr. Ende Juli 1941
27. L.S. (38) + 6 Bodenzerstör.

Rollwage, Herbert (rechts) JG 53 Bf 109F
Comiso/Sizilien 8. Aug. 1942
30. L.S. (102)

Rollwage, Herbert
5./JG 53
Bf 109 G
Trapani/Italien
10. Juli 1943
47. L.S. (102)

Rudorffer, Erich
Abbeville/Frankr.
40. L.S. (222)

St. Kap. 6./JG 2
8. Dez. 1941

Bf 109 F

Rudorffer, Erich
St. Kap. 6./JG 2
Fw 190 A
Beaumont/Frankr.
19. Aug. 1942
45. L.S. (222)

Russ, Otto (rechts) | 4./JG 53 | Bf 109 G
Eschborn b/Frankf. Main | März 1944
15. L.S. (27) | links: Stefan Litjens

Sayn-Wittgenstein, Heinrich Prinz zu (Mitte) | Ju 88 C
Ostfront | Kdr. I./NJG 100
29. L.S. (83) | Mai 1943

Schauder, Paul	Stab/JG 26	Bf 109 F
St. Omer/Frankr.	Sommer 1942	
13. L.S. (20)		

Schenck, Wolfgang	St. Kap. 1./SKG 210	Bf 110E
Seschtinskaja/Ostfr.	Sept. 1941	
15. L.S. (18) + 5 Panzer		

Schieß, Franz — Stab/JG 53 — Bf 109 F
St. Omer/Frankr. — 13. Apr. 1941
1. L.S. (67)

Schieß, Franz — Stab/JG 53 — Bf 109 F
Biala-Zerkow/Ostfr. — 29. Juli 1941
14. L.S. (67)

Schieß, Franz
Comiso/Sizilien
17. L.S. (67)

Stab/JG 53
15. Juni 1942

Bf 109 F

Schieß, Franz
Stab/JG 53
Bf 109 G
El Aouina/Tunis
29. Jan. 1943
37. L.S. (67)

Schmidt, Erich Surash/Ostfr. 40. L.S. (47)	III./JG 53 7. Aug. 1941 links: 1. Wart	Bf 109 F

Schmidt, Erich (stehend) Surash/Ostfr. 44. L.S. (47)	III./JG 53 Aug. 1941	Bf 109 F

Schmidt, Erich Surash/Ostfr. 47. L.S. (47)	III./JG 53 29. Aug. 1941	Bf 109 F

Schnaufer, Heinz-Wolfg. St. Trond/Belgien 47. L.S. (121)	St. Kap. 12./NJG 1 15. Febr. 1944	Bf 110G

Schnaufer, Heinz-Wolfg. (Mitte) Kdr. IV./NJG 1 Bf 110 G
St. Trond/Belgien 9. Okt. 1944
100. L.S. (121) links: Bordf. Rumpelhardt rechts: Bordsch. Gänsler

Schnaufer, Heinz-Wolfg.
Kdore. NJG 4
Bf 110 G
Imp. War Museum, London
letzter L.S. 7. März 1945
121. L.S. (121)

Schnell, Siegfried | St. Kap. 9./JG 2 | Bf 109F
St. Pol/Frankr. | Juli 1941
44. L.S. (93)

Schnell, Siegfried | St. Kap. 9./JG 2 | Bf 109 F
Théville/Frankr. | Mai 1942
52. L.S. (93)

Schnell, Siegfried
St. Kap. 9./JG 2
Bf 109 F
Théville/Frankr.
Ende Mai 1942
57. L.S. (93)

Schnell, Siegfried
St. Kap. 9./JG 2
Fw 190 A
Théville/Frankr.
Anfang Juni 1942
62. L.S. (93)

Schnell, Siegfried
Vannes/Frankr.
75. L.S. (93)

St. Kap. 9./JG 2
18. Febr. 1943

Fw 190A

Schöpfel, Gerhard Coquelles/Frankr. 22. L.S. (40)	Kdr. III./JG 26 Okt. 1940	Bf 109E

Schramm, Herbert Lepel-Ost/Ostfr. 18. L.S. (42)	III./JG 53 6. Juli 1941	Bf 109 F

Schroer, Werner
Kdr. II./JG 27
Bf 109 G
Luftwaffenmuseum Uetersen
Stand: 11. Jan. 1944
90. L.S. (114)

Schubert, Hans
3./JG 1
Bf 109 E
de Kooy/Holland
28. Mai 1941
4. L.S. (8)

Schuhmacher, Leo	2./ZG 76	Bf 110 C
Stavanger/Norwegen	15. Aug. 1940	
4. L.S. (23)		

Schulz, Albert (2. v. links)	I./NJG 2	Do 17Z-10
Gilze Rijen/Holland	1941	
3. L.S. (Ges. ?)		

Schultz, Otto | 4./JG 51 | Bf 109F
Star-Bychow/Ostfr. | 26. Juli 1941
11. L.S. (73)

Seckel, Georg | III./JG 77 | Bf 109 F
Jassy/Rumänien | 4. Juli 1941
6. L.S. (40)

Seegatz, Hermannn
Petsamo/Finnland
27. L.S. (31)

St. Kap. 8./JG 5
Frühjahr 1941

Bf 109 E

Seifert, Johannes
St. Omer/Frankr.
34. L.S. (57)

Kdr. I./JG 26
17. Mai 1942

Fw 190A

Seiler, Reinhard
St. Kap. 1./JG 54
Bf 109 F
Mal-Owsischtschi/Ostfr.
16. Aug. 1941
21. L.S. (109)

Setz, Heinrich
Sarabush/Krim
45. L.S. (138)
St. Kap. 4./JG 77
März 1942
Bf 109 E

Setz, Heinrich St. Kap. 4./JG 77 Bf 109 F
Sarabush/Krim 9. Mai 1942
67. L.S. (138)

Sinner, Rudi I./JG 27 Bf 109F
Tmimi/Nordafrika 26. Juni 1942
8. L.S. (39)

Späte, Wolfgang	St. Kap. 5./JG 54	Kl 35 D
Staraja-Russa/Ostfr.	7. Okt. 1941	
47. L.S. (99)	Reiseflugzeug	(Pers. courier aircraft)

Specht, Günther	Kdr. II./JG 11	Bf 109G
Quakenbrück	29. Nov. 1943	
22. L.S. (mind. 32)		

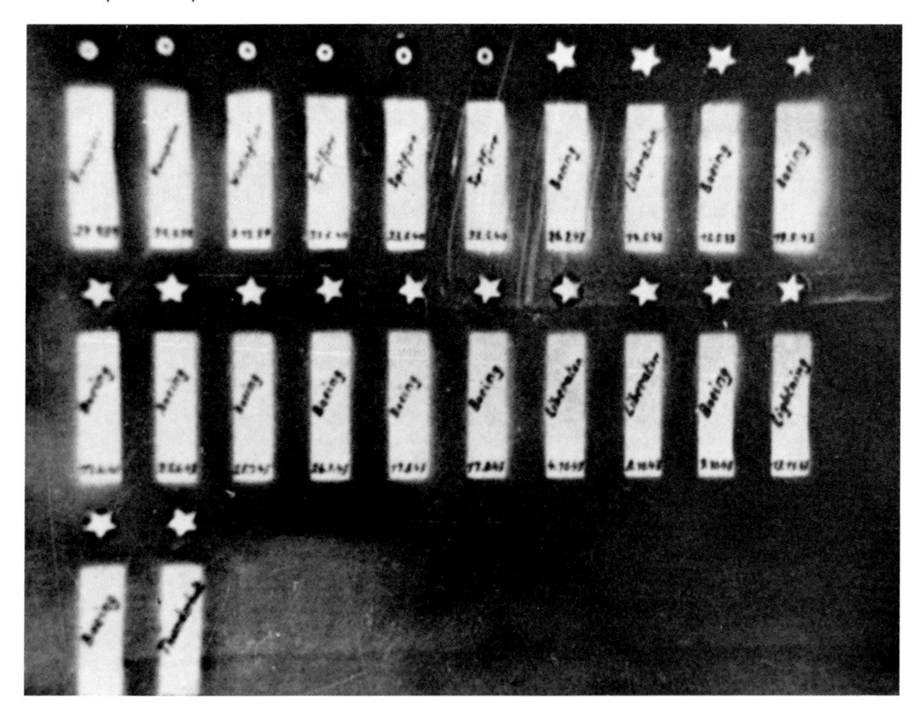

Specht, Günther (links)
Wunstorf
30. L.S. (mind. 32)

Kdr. II./JG 11
5. Jan. 1944
rechts: Dipl. Ing. Kurt Tank

Bf 109 G

Spies, Wilhelm Sarudinje/Ostfr. 19. L.S. (21) Kdr. I./ZG 26 Sommer 1941 Bf 110 D

Spies, Wilhelm
Kdr. I./ZG 26
Bf 110 E
Siwerskaja/Ostfr.
Herbst 1941
21. L.S. (21)
abgebildet: Bordfunker

Stahlschmidt, Hans Arnold
St. Kap. 2./JG 27
Bf 109 F
Quotaifya/Nordafr.
23. Aug. 1942
48. L.S. (59)
+ 1 Schiff

Steiß, Heinrich
St. Dizier/Frankr.
17. L.S. (21)

II./JG 27
Sept. 1943

Bf 109 F

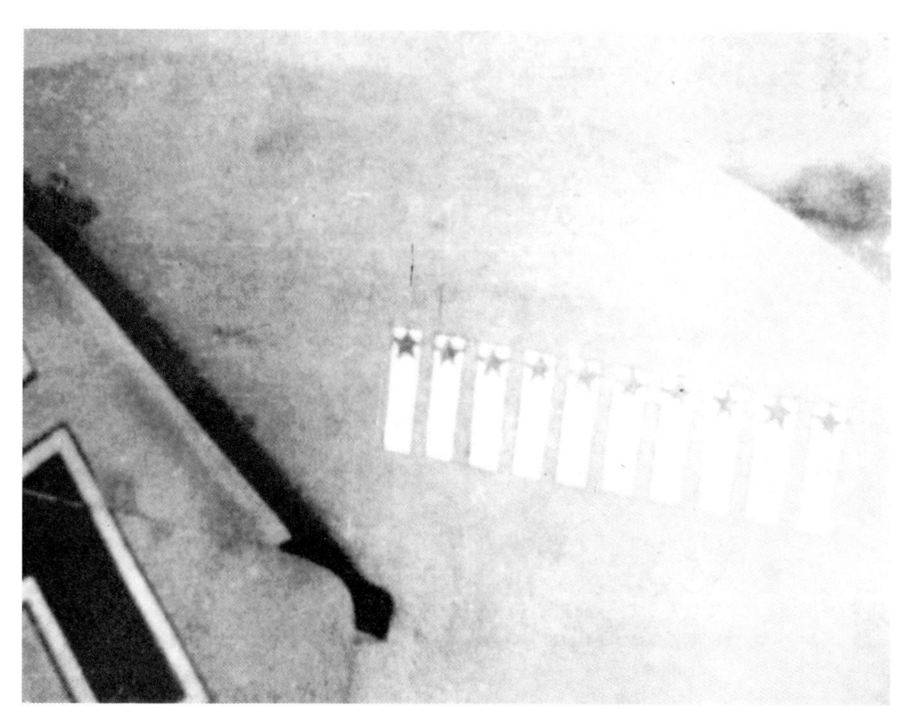

Stollnberger, Hans Ostfront 10. L.S. (Ges. ?)	8./SG 1 Sommer 1942	Bf 109 F
Stolte, Paul de Kooy/Holland 2. L.S. (43)	St. Kap. 3./JG 1 1941	Bf 109 E

Stückler, Alfred (links) — III./JG 27 — Bf 109 F
Rhodos/Ägäis — 13. Juni 1943
4. L.S. (10)

Tonne, Günther — II./SKG 210 — Bf 110F
Seschtinskaja/Ostfr. — Okt. 1941
15. L.S. (mind. 20)

Tonne, Wolfgang vor Stalingrad 95. L.S. (122)	St. Kap. 3./JG 53 Sept. 1942	Bf 109F

Tietzen, Horst Desvres/Frankr. 11. L.S. (27)	5./JG 51 29. Juli 1940	Bf 109 E

Tietzen, Horst	5./JG 51	Bf 109E
Marquise/Frankr.	15. Aug. 1940	
17. L.S. (27)		

Vinke, Heinz
11../NJG 1
Bf 110 F
Leeuwarden/Holland
26. Mai 1943
25. L.S. (54)

Weiß, Robert
I./JG 54
Bf 109 G
Wesenberg/Baltikum
März 1944
73. L.S. (121)

Weißenberger, Theodor
10. (Z)/JG 5
Bf 110 E
Kirkenes/Norwegen
15. Mai 1942
20. L.S. (208)

Welter, Kurt II./NJG 11 Bf 109K
Jüterbog Okt. 1944
33. L.S. (51)

Werra, Franz von Adj. II./JG 3 Bf 109E
Marsden/England nach Notlandung 5. 9. 1940
8. L.S. (21) + 5 Bodenzerstör.

Wick, Helmut (3. von links) Mardyck/Belgien 21. L.S. (56)	St. Kap. 3./JG 2 30. Aug. 1940	Bf 109E

Wick, Helmut Mardyck/Belgien 42. L.S. (56)	Kdr. I./JG 2 6. Okt. 1940	Bf 109E

Wick, Helmut — Kdore. JG 2 — Bf 109E
Beaumont/Frankr. — 9. Nov. 1940
54. L.S. (56)

Willius, Karl — III./JG 51 — Bf 109E
St. Omer/Frankr. — Herbst 1940
3. L.S. (50)

Willius, Karl — III./JG 51 — Bf 109F
Orscha/Ostfr. — Ende Juli 1941
12. L.S. (50)

Wurmheller, Josef (links) — St. Kap. 9./JG 2 — Fw 190A
Beaumont/Frankr. — Sept. 1943
81. L.S. (102)

Wurmheller, Josef
Beaumont/Frankr.
74. L.S. (102)

St. Kap. 9./JG 2
Juli 1943

Fw 190A

Wurmheller, Josef Beaumont/Frankr. ca. 90. L.S. (102)	St. Kap. 9./JG 2 Frühjahr 1944	Fw 190A

Lützow, Günther Schatalowka/Ostfr.	Kdore. JG 3 10. Okt. 1941	Bf 109F

Unfortunately the authors of this book have not been able to identify correctly each of the many available pictures of rudder markings, some of them remarkable enough to warrant inclusion in this book. Some reader might be able to clarify some details on the identity of the respective pilots. In this case please contact the publisher. Out of context is the first picture of this series showing phony rudder markings applied to an He 100D that never fired a shot in anger.

Leider war es den Verfassern des Buches nicht möglich, alle ihnen zur Verfügung stehenden Ruderfotos einwandfrei zu identifizieren, jedoch sind auch die Markierungen an unbekannten Flugzeugen so bemerkenswert, daß wir auf deren Wiedergabe nicht verzichten möchten. Möglicherweise ergeben sich aus dem Leserkreis Hinweise, die zur Ermittlung der Flugzeugführer solcher, noch zu klärender Markierungen beitragen können. Angaben die weiterhelfen könnten bitte deshalb an einen der Autoren über den Hoffmann-Verlag. Aus dem Rahmen der echten Erfolgsbuchungen fällt das erste Foto dieser Reihe, das Leitwerk einer He 100D; hier wurden zu reinen Propagandazwecken Abschüsse aufgemalt, die nie mit dieser Maschine errungen worden waren, einfach, weil mit diesem Baumuster keine »scharfen« Einsätze mit Feindberührung geflogen wurden.

He 100D, Propaganda-Aufnahme. Die aufgemalten Abschüsse wurden in Wirklichkeit nicht erzielt.

A propaganda picture of an He 100D with phony »kill« bars being applied, the aircraft never being used in actual combat.

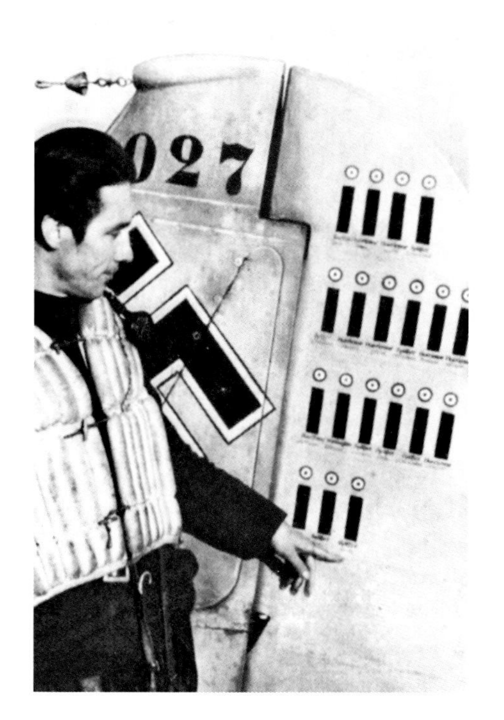

Unbekannt
Fw 190 A
W. Nr. 027
5./JG . . .
Frankreich 1941
19 Luftsiege

Unbekannt
Einheit:
Fw 190 A
54 Luftsiege

Unbekannt
Fw 190 A
W. Nr. 2463
Westen oder Reichsgebiet
7 Luftsiege

Unbekannt
Ju 88 C
südl. Ostfront
22 Luftsiege

Rudolf Schönert?

Unbekannt
II./JG 77?
Winter 1939/40
3 Luftsiege

Bf 109E W.Nr. 1279
Norddeutscher Raum

Unbekannt
4./JG 52
19. August 1940
5 Luftsiege + 1 Beobachtungsballon

Johannes Steinhoff?
Bf 109E W. Nr. 5256

Unbekannt
III./JG 2
Bf 109 F
Frankreich 1940/41
20 Luftsiege

Unbekannt
Kdr. I./JG . . .
Brest/Frankreich
15 Luftsiege

Bf 109E
1941

Unbekannt
JG 26?
Bf 109
Frankreich 1940/41
6 Luftsiege

Unbekannt
II./JG 26?
Frankreich 1940
14 Luftsiege

Bf 109E

Unbekannt
JG 52
Bf 109 E
Ort:
7 Luftsiege

Unbekannt Bf 109E W. Nr. 6389
1. (J)/LG 2
Rumänien
10 Luftsiege

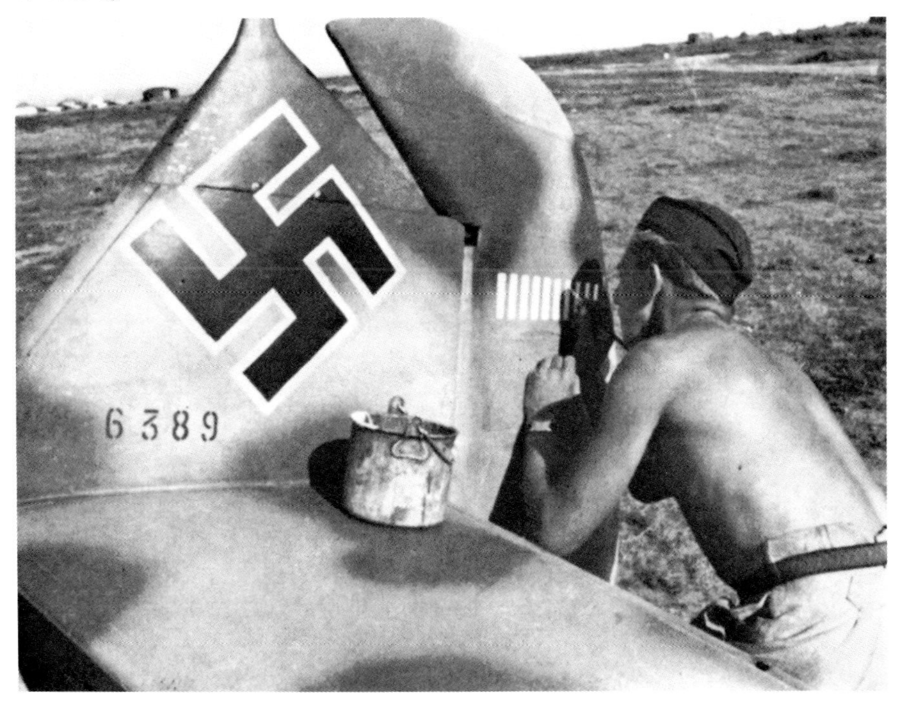

Unbekannt Bf 109F W. Nr. 7205
Einheit?
54 Luftsiege

Unbekannt Bf 109F
Kdore. JG 77?
Ostfront
4 bestätigte + 3 noch zu bestätigende Luftsiege

Unbekannt	Bf 109F
II./JG 3	Aug. 1941
Stschastliwaja/Ostfr.	
15 Luftsiege + 2 Schiffe	

Unbekannt	Bf 109F
bei Solugubowka/Ostfr.	29. Sept. 1941
10 Luftsiege	

Unbekannt Bf 109
Einheit: 16. Apr. 1942
Ort:
8 Luftsiege

Unbekannt Bf 109F
Einheit:
Ort:
19 Luftsiege rechts: Ltn. z. See Rudolf Porath

Unbekannt Bf 109G W, Nr. 15912
Einheit:
Ort:
4 Luftsiege

Unbekannt Bf 110 W. Nr. 4293
ZG 26?
Ort:
6 Luftsiege links: Ltn. z. See Rudolf Porath

Unbekannt
III./ZG 26
Bf 110 D W. Nr. 3412
Ort:
8 Luftsiege

Unbekannt Bf 110
Einheit:
Ort:
9 Luftsiege + 3 Sperrballone

Unbekannt
NJG
Bf 110
Ort:
3 Luftsiege

Glück im Unglück hatte dieser Sergeant der RAF, der am 18. Dez. 1939 aus einer brennenden »Wellington« mit dem Fallschirm absprang, und hier im Kasino der I./ZG 76 zu Gast ist.

On 18. Dec. 1939 this RAF Sergeant of a »Wellington« crew parachuted to safety from his burning aircraft. Here he is entertained by personnel of the I./ZG 76.

The following series of pictures differs from the previous one in showing not the individual score but the unit score on vehicles, score charts at the unit's headquarters, etc.. In addition, some fighter pilots display their elaborately carved »victory sticks« that were, similar to a mascot, carried on board of the aircraft when moving to a new location. Finally, this chapter gives an idea of the reception given to a fighter pilot bringing home an »anniversary« victory, either a personal one or one for the unit.

Die nachstehende Fotoreihe zeigt im Gegensatz zu den vorangegangenen Bildern nicht die indivduelle, sondern die pauschale Markierung von Abschüssen ganzer Einheiten auf Fahrzeugen, Abschußtafeln in Gefechtsständen, etc. Daneben präsentieren einige Jagdflieger ihre kunstvoll geschnitzten Abschußstöcke, die gleich einem Maskottchen bei jeder Verlegung in die Maschine gepackt wurden, und so auf dem Luftweg den neuen Einsatzhafen erreichten. Schließlich vermittelt dieses Kapitel auch einen Eindruck vom Empfang solcher Jagdflieger, die einen Jubiläumsabschuß, sei es zu einem eigenen, oder zum Jubiläum einer Einheit mit nach Hause brachten.

Die Ruder von Wilhelm Balthasar (links) und Helmut Wick (rechts) in der Offiziersmesse des JG 2 »Richthofen«.

The vertical rudders of Wilhelm Balthasar (left) and Helmut Wick (right) in the officers' mess of the JG 2 »Richthofen«.

Der Gefechtsanhänger der 7./JG 54 nach dem 200. Luftsieg der Staffel auf dem Platz von Siwerskaja/Ostfr. im Juni 1942.

The operations trailer of the 7./JG 54 after the 200th squadron victory on the airfield of Siverskaya, Russia, in June 1942.

Im August 1941 zeigt die Bemalung des Staffelwagens der 9./JG 54 die 76 Abschüsse dieser Staffel auf. Sarodinje/Ostfront.

In August 1941 a squadron vehicle of the 9./JG 54 depicts this squadron's 76 victories. Sarodinye, Russia.

Klaus Bretschneider,
Staff. Kap. der 5./JG 300 steht mit seinem Abschußstock vor seiner Fw 190A »Rauhbautz VII«. Stand Sommer 1944 nach 21 Luftsiegen (Ges. 31)

Klaus Bretschneider, Staffelkapitaen (Squadron Leader) of the 5./JG 300 with his »victory stick« next to his Fw 190 A »Rauhbautz VII« (free translation: Tough Guy). Summer 1944 after achieving 21 of his total of 31 »kills«.

Johannes Wiese (links, mit Abschußstock) als Staffelkapitän der 2./JG 52 zusammen mit Gen. Oberst Stumpff in Rossosh/Ostfr., Jan. 1943. Wiese erzielte insgesamt 133 Ostabschüsse.

Johannes Wiese (left, with »victory stick«) as Staffelkapitaen of the 2./JG 52 together with Generaloberst Stumpff in Rossosh, Russia, January 1943. Wiese achieved a total of 133 aerial victories in the East.

Ruder und Abschußstöcke einer unbekannten Einheit an der Ostfront.

Rudder and »victory sticks« of an unknown unit in the East.

Abschußtafel mit 90 Luftsiegen der 5. Staffel/JG 52 vor einer Bf 109 F-2 auf einem russischen Flugplatz.

»Kill« scoreboard with 90 aerial victories of the 5th Staffel/JG 52 in front of a Bf 109 F-2 on a Russian airfield.

Nach der Landung wird dem Kdore. des JG 2, Kurt Bühligen zu seinem 100. Luftsieg gratuliert. 9 Juni 1944, Westfront.

After landing the Kommodore of the JG 2, Kurt Buehligen, is being congratulated for his 100th aerial victory. 9 June 1944, western front.

Großer Blumenstrauß und Jubiläums-Plakat für Walter Oesau nach seinem 100. Abschuß am 26. Okt. 1941. Der Kdore. des JG 2 in seiner Unterkunft in St. Pol/Frankr.

A bunch of flowers and an anniversary poster for Walter Oesau after his 100th »kill« on 26 October 1941. The picture shows the Kommodore of the JG 2 in his lodgings in St. Pol, France.

Am 26. Jan. 1943 feiert »Assi« Hahn (links) seinen 100., Max Stotz seinen 150. Luftsieg bei der II./JG 54 in Rjelbitzy/Ostfr. Hans Beißwenger (rechts) läßt sich dabei zum Leeren der »Tassen« nicht zweimal einladen.

On 26 January 1943 »Assi« Hahn (left) is celebrating his 100th, Max Stotz his 150 th aerial victory with the II./JG 54 at Ryelbitzy, eastern front. Hans Beisswenger (right) does not have to be asked twice to help emptying the »coffee« cups.

Walter Nowotny erzielte den 300. Luftsieg der 1./JG 54. Sein Empfang nach der Landung in Gatschine/Ostfront im November 1942.

Walter Nowotny achieved the 300th aerial victory of the 1./JG 54. The picture was taken on occasion of his welcome after his return at Gatchine, eastern front, in November 1942.

»Großer Bahnhof« für Wilhelm Crinius, der den 1000. Gegner für die I./JG 53 Ende August 1942 vor Stalingrad bezwingt.

»Red carpet« treatment for Wilhelm Crinius who shot down the 1000th enemy of the I./JG 53 in late August 1942 near Stalingrad.

Dem Schützen des 2000. Luftsieges der II./JG 3, Uffz. Walter Stienhaus wird am 12. Juli 1943 dieses Diplom in Charkow ausgestellt.

The marksman who won the 2000th aerial victory for the II./JG 3, Unteroffizier Walter Stienhaus, was awarded this certificate in Charkov on 12 Juli 1943.

7000. Abschußerfolg des JG 54 durch Albin Wolf, gleichzeitig sein 135. Erfolg. Petzeri/Ostfr. am 23. März 1944.

The 7000th victory for the JG 54 was also Albin Wolf's 135th personal »kill«. Petzeri, eastern front, 23 March 1944.

Kommodore Hrabak gratuliert Adolf Borchers zum 10 000. Luftsieg des JG 52. Borchers war zu dieser Zeit Kdr. I./JG 52 und hatte 118 Gegner auf seinem Konto.

The Kommandeur of the I./JG 52, Adolf Borchers, is congratulated for the 10 000th aerial victory of the JG 52, also his own 118th »kill«, by Kommodore Hrabak.

2000. Feindflug der 4. (F)/33 mit Bf 110 G, Ostfront.

2000th mission of the 4. (F)/33 in the East, the picture showing a sign in front of a Bf 110 G.

Urkunde für die Besatzung des 1000. Feindfluges der 3. (H)/21 Ostfront 5. 6. 1942

Certificate for the crew that flew the 1000th mission of the 3. (H)/21. Eastern front, 5 June 1942.

Do 217 E des KG 40 mit Markierung der Einsatzräume.
A Do 217E of the KG 40, the markings showing the unit's various operational locations.

He 111H der 5./KG 55 mit Staffelzeichen auf dem Seitenruder. Ostfront.
An He 111H of the 5./KG 55 with squadron symbol on the rudder. Eastern front.

2. Various Kinds of Success Scores

As told in the previous chapter, the German Air Force started keeping score of »kills«, primarily aerial victories, on its aircraft with its participation in the Spanish Civil War. With the beginning of the 2nd World War on 1. September 1939 not only fighter and destroyer units operated on a large scale but bomber, dive bomber and reconnaissance units as well. On a number of occasions the crew of such a bomber or reconnaissance aircraft succeeded in shooting down an enemy attacker and observing its impact on the ground. As with the fighters, so the crews of other units too were submitting their defensive »kill« claims to the RLM for confirmation, which, provided there were the necessary witnesses available, was given in due time. What the fighters could do, the crews of the KG's, St. G's, (F) Staffeln and coastal patrol aircraft could do as well. Thus »kill bars« began to appear on the rudders and fuselages of aircraft like the He 111, Do 17, Ar 196, Ju 88 and Do 217 the same way as previously on Bf 109 and Bf 110 aircraft, to prove to the colleagues of the »speedy faculty« that they were no the only ones who could handle their guns effectively.

As a result of the rapidly mounting number of missions, according to location and necessity up to 5 missions a day, the various crews were aiming for the different grades of the »Frontflugspange« (Mission Bar), as outlined in the regulation dated 10 February 1941. In order to keep the number of missions flown not confined to the pages of the flight log here too a sort of scorekeeping on the rudder was introduced. This custom, however, lasted only a short time since the different aircraft in a squadron were not always flown by the same crews and a continuous marking was thus made very difficult. Usually a dot or lateral bar was painted on the vertical stabilizer for each mission flown. Upon awarding of the »Frontflugspange« this medal was painted on instead of the hitherto flown missions and new missions were added as before (as with Do 217, Ju 88).

Very similar to the fighter units' anniversary »kills« the bomber, dive bomber, reconnaissance, transport and coastal patrol units were celebrating their mission anniversaries and the respective crew received a certificate.

2. Erfolgsbuchungen verschiedenster Art

Wie aus dem Vorstehenden bereits zu entnehmen, bürgert es sich bei der Deutschen Fliegertruppe erst mit der Beteiligung am Spanischen Bürgerkrieg ein, Erfolge, in erster Linie Abschußerfolge auf den Flugzeugen festzuhalten.

Der Ausbruch des 2. Weltkrieges am 1. 9. 1939 sieht jedoch von Anbeginn an nicht nur Jagd- und Zerstörerverbände, sondern in weit größerem Rahmen die Kampf-, Sturzkampf- und Aufklärungseinheiten der Luftwaffe im Einsatz. Nicht selten gelingt es dabei einer Besatzung während eines Bomben- oder Aufklärungsfluges einen gegnerischen Angreifer abzuschießen und den Aufschlag zu beobachten. Wie bei den Jägern werden auch von den Besatzungen anderer Verbände solche Defensiv-Abschüsse beim RLM zur Bestätigung eingereicht, und, wenn die notwendigen Abschußzeugen beigebracht werden können, auch von den Kommissionen anerkannt. Was nun den Jagdfliegern recht, ist den Besatzungen der KG's, St. G.'s, (F)-Staffeln und Küstenfliegern billig. Wie auf den Rudern der Bf 109 und Bf 110, tauchen nun auch Abschußbalken auf Rudern und Rümpfen von He 111, Do 17, Ar 196, Ju 88, Do 217 usw. auf und zeigen damit den Kollegen von der »schnellen Fakultät« an, daß sie nicht allein mit ihren MG's umzugehen verstehen.

Mit der rapide anwachsenden Zahl von Feindflügen, je nach Liegeplatz und den Erfordernissen der Frontlage bis zu 5 an einem Tage, steuern die eingesetzten Besatzungen die Verleihung der verschiedenen Stufen der Frontflugspange (Bestimm. vom 10. 2. 1941) an. Um nun seine Feindflugzahl nicht nur im Flugbuch vorweisen zu können, wird auch hier eine Buchhaltung der geflogenen Einsätze auf dem Ruder der eigenen Maschine eingeführt. Diese Gewohnheit hält sich jedoch nur für kurze Zeit, da die Staffelmaschinen nicht immer von den gleichen Besatzungen geflogen werden, und so eine kontinuierliche Markierung erschwert wird. Gewöhnlich wird je Feindflug auf der Seitenruderflosse des Flugzeuges ein Punkt oder Querbalken aufgemalt, bei Verleihung der Frontflugspange anstelle der bisherigen Feindflugzahl das Symbol für diese Auszeichnung gesetzt, und die danach folgenden Feindflüge als Punkte weiter markiert (s. Do 217, Ju 88).

Ganz ähnlich wie bei den Jagdverbänden eine Jubiläumsabschußzahl, wird bei Kampf-, Stuka-, Aufklärungs-, Transport und Seefliegereinheiten ein Feindflugjubiläum besonders gefeiert und der Jubiläumsbesatzung ein besonderes Diplom verliehen. Auch die Flieger der alliierten Kampfverbände verbuchen die Zahl ihrer »missions« auf den Maschinen, mit denen sie ihre Einsätze heil überstanden haben.

The airmen of the Allied bomber units were also keeping book of their missions on the aircraft they succeeded in coming back safely. These markings were found mostly as bomb silhouettes on the aircraft nose or side underneath the cockpit. Many British and American bombers showed a number of swastikas or crossed bars in addition to their mission markings each of these standing for a downed German aircraft. These »kill« markings were not to be taken too literally since bomber crews were almost never able to prove an actual »kill« beyond any doubt (observed impact, explosion in the air, etc.).

As a rule every single gunner in a bomber formation who had fired at an attacking fighter going into a dive, showing some »smoke« or only some slightly uncontrolled attitude, naturally reported this as his personal »kill«.

In addition to the marking of the number of missions on their aircraft the German bomber crews for a time practiced designating their targets as well. The squadron painters were doing great things this way and in no time transformed the tailplanes of an aircraft into a panorama complete with factories, tent encampments, aircraft hangars and entire maps. But in the long run this kind of artistry may have become too time-consuming or the available space was not large enough, in any case were such markings no longer found in the final stages of the 1941 air offensive against Britain. For a considerable time, however, the rows of ship silhouettes were kept on the rudders of bombers that succeeded in sinking enemy ships (Do 217, He 111, He 115, Ju 88). Normally they belonged to those units specialising in anti-shipping raids, such as KG 25, KG 26, KG 28, KG 30, KG 40, KG 77, K.Gr. 126, K.Gr. 806 and several torpedo-carrying coastal patrol units. In almost all cases the date of the sinking and the estimated ship's size in tons was listed in addition to the ship's outline. Without doubt such estimates were often made in a very generous way and »rounded up« accordingly.

Often, too, a heavily damaged freighter was reported as sunk that managed to reach port after all. In most cases such reported sinkings, provided they were not observed during the raid without a doubt and corroborated by witnesses, were to be confirmed later by visual or photo reconnaissance due to the limited time of the raid.

Hier findet man die Feindflugmarkierungen meist als Bombensilhouetten am Rumpfbug oder unterhalb des Besatzungsraumes auf den Rumpfseiten. Auf vielen Bombenmaschinen englischer und amerikanischer Verbände sind neben diesen »missionmarkings« noch Reihen von Haken- oder Balkenkreuzen zu finden, wobei je ein solches Nationalitätszeichen für den Abschuß eines deutschen Flugzeuges steht. Diesen Abschußmakierungen ist allerdings insofern nur sehr wenig Bedeutung beizumessen, als eine Bomberbesatzung in den allerwenigsten Fällen den tatsächlichen Abschuß (beobachteter Aufschlag, Explosion in der Luft etc.) nachweisen konnte. In der Regel meldete jeder Bordschütze eines Bomberverbandes, der auf einen angreifenden Jäger gehalten hatte, nachdem dieser »kokelnd« nach unten weggestürzt war, oder eine nur annähernd unkontrollierte Fluglage gezeigt hatte einen Abschuß.

Neben der Buchung der Feindflugzahl auf den Flugzeugen wird bei den deutschen Kampfverbänden für kurze Zeit auch das Kennzeichnen des angegriffenen Zieles praktiziert. Die Staffelmaler leisten auf diesem Gebiet beachtliches und verwandeln das Ruder einer Einsatzmaschine im Nu in ein Panorama mit Fabrikanlagen, Zeltlagern, Flugzeughangars und ganzen Landkarten. Auf die Dauer wurde diese Art der Malerei wohl doch zu zeitraubend, oder die Ruder reichten für weitere Zieldarstellungen nicht mehr aus; jedenfalls sind solche Markierungen in der Endphase der Luftoffensive gegen England 1941 schon nicht mehr anzutreffen.

Sehr lange dagegen halten sich die Reihen von Silhouetten versenkter gegnerischer Schiffe auf den Rudern der Kampfmaschinen (Do 217, He 111, He 115, Ju 88). In der Regel handelt es sich hierbei um Flugzeuge solcher Verbände, die auf die Bekämpfung der gegnerischen Schiffahrt spezialisiert sind, wie: KG 25, KG 26, KG 28, KG 30, KG 40, KG 77, KGr. 126, KGr. 806 und verschiedene Küstenfliegergruppen (torpedotragend). Fast immer wird neben der Schiffssilhouette das Datum der Versenkung oft dazu mit der geschätzten BRT-Zahl des versenkten Fahrzeuges angegeben. Zweifelsohne wird bei solchen Schätzungen der Schiffsgröße sehr großzügig »über den Daumen gepeilt« und nach oben abgerundet, möglicherweise auch mancher nur schwer getroffene Frachter, der sich doch noch in einen Hafen schleppen kann als versenkt angenommen.

Oft werden solche Versenkungsmeldungen, sofern der Untergang des Schiffes nicht schon während des Angriffes einwandfrei durch Zeugen beobachtet werden konnte, wegen der Kürze der Angriffszeit erst aufgrund von später geflogener Bild- oder Augenerkundung bestätigt.

From 1943 on a number of Ju 52's with huge rings, suspended under fuselage and wings, were treated with much respect when they appeared on airfields in the South or the Balkan since they displayed an astonishing number of »kill bars« on their tailplanes. Obviously the gunners, handling their single weapons in a fantastic way, had to be genuine William Tells to score such remarkable successes. Alas, these rudder marks showed no aerial victories but magnetic mines that had been brought to explosion by these low-flying »Auntie Ju's« with the help of their curious rings, generating a powerful magnetic field.

As a final special marking the displaying of the various theaters of war may be mentioned in which crews and aircraft had been operating. Normally painted underneath the cockpit were the roundels of the various enemies missions had been flown against. This kind of markings were comparatively rare primarily found on unit leaders' aircraft, never on all aircraft of a squadron or a group.

Voller Ehrfurcht werden ab 1943 im Süd- und Südostraum einige Ju 52 mit einem riesigen, unter Rumpf und Flächen aufgehängten, flachen Ring und einer erklecklichen Anzahl weißer Balken an den Rudern, auf den Flugplätzen dieser Einsatzgebiete bestaunt. Die Schützen im jeweils einzigen Abwehrstand dieser Ju 52 müßten mit ihrer »Kugelspritze« wahre Wilhelm Tells sein, um eine solch große Anzahl von Feindflugzeugen zur Strecke gebracht zu haben. Es handelt sich bei diesen Strichen am Leitwerk jedoch keineswegs um Abschüsse, vielmehr sind diese Ju 52 mit dem ominösen untergehängten Ring, mit dessen Hilfe ein elektromagnetisches Feld erzeugt wird in Minensuchstaffeln der Luftwaffe im Einsatz. Beim niedrigen Überfliegen von Magnetminen in Hafeneinfahrten und Flußläufen, werden die Minen zur Detonation gebracht, und der »Tante Ju« für jeden Räumerfolg ein weißer Balken auf das Ruder gemalt.

Als letzte Sondermarkierung sei an dieser Stelle noch die Kennzeichnung von Kriegsschauplätzen, auf denen die Besatzungen mit ihren Flugzeugen eingesetzt waren erwähnt. Meist unterhalb der Kabine aufgemalt, finden sich die Kokarden der Gegner, gegen die die Einheiten ihre Angriffe flogen. Diese Art der Bemalung ist relativ selten anzutreffen, vorwiegend auf dem Flugzeug eines Verbandsführers, nie auf allen Maschinen einer Staffel oder gar Gruppe.

Arado Ar 95W
Einheit:
Ort:
1 Luftsieg am 19. 8. 1941

Arado Ar 95W
Einheit:
Ort:
1 U-Boot-Versenkung am 11. 8. 1941

Arado Ar 196 A W. Nr. 196 0056
Einheit:
Ort:
1 Schiffs-Versenkung

Dornier Do 17 P 4U + LL
3. (F)/123
Ort:
1 Luftsieg 8. Sept. 1939

Dornier Do 217E
Besatz.: Albert Hain
KG 2
Ort: Frankreich
Schiffsversenkungen

Dornier Do 217E
Besatzung:
Stab II./KG 40
Ort: Frankreich
2 Abschüsse
Industrieziele
3 Sperrballone

Do 217E F8+DP 6./KG 40
Besatz.:
Ort: Frankreich
39 Feindflugmarkierungen

Do 217E KG 40
Besatz.:
Frankreich
Frontflugspange für Kampfflieger

Fw 200 C — F8 + BW
Besatz.: Edmund Daser — 12./KG 40
Frankreich — 1941
Schiffsversenkungen + 45 Feindflüge

Fw 200C
Bes.: Buchholz?
I./KG 40
Südfrankreich
1941
Schiffsversenkungen + Feindflüge

He 111H W Nr. 7186
Bes.:
KG 53?
Ostfront
12 Luftsiege

He 111H
Bes.:
KG 55
Frankreich
1941
Industrieziele
+ Flugplätze
+ Truppenlager

He 111H
Bes.:
Einheit:
Ort:
Schiffsversenkungen

He 111H
Bes.:
Ort: Norwegen?
Schiffsversenkungen

W. Nr. 7128
KG 26
Sommer 1942

He 115B K6 + LH
Bes.:
1./Kü. Fl. Gr. 406
Norwegen 1941
Schiffsversenkungen
+ 1 Luftsieg

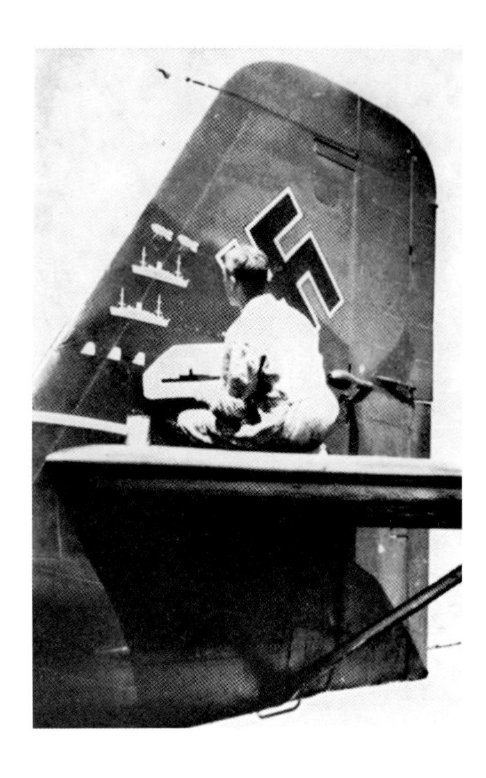

He 115B
Bes.:
Einheit:
Frankreich
2 Sperrballone
+ 2 Schiffe
+ 3 Luftsiege

Hs 129B
Rudolf-Heinz Ruffer
Kuban/Ostfr.
8 Panzer (Ges. 72 Panzer)

W. Nr. 0364
8. (Pz.)/SG 1
8. 4. 1943

Hs 129 B
Rudolf-Heinz Ruffer
St. Kap. 8. (Pz)/SG 1
Kuban/Ostfr.
Mai 1943
13 Panzer (Ges. 72 Panzer)

Ju 52/3mg6e/MS
Bes.:
Ort:
28 Minenräumerfolge

W. Nr. 3286
Minensuchstaffel der L.W.

Ju 88A W. Nr. 1333
Bes.:
I./KG 30
Ort:
20 Feindflüge

Ju 88A W, Nr. 1726
Bes.:
I./KG 30
Ort:
1 Luftsieg
+ 2 Sperrballone
+ 18 Feindflüge

Ju 88A — 7./KG 1
Bes.:
Ort:
Feindflüge + Frontflugspange für Kampfflieger

Ju 88A — 4D + MR
Bes.: — 7./KG 30
Mittelmeerraum
Schiffsversenkungen

Ju 88A 3Z + DB
W. Nr. 1016
Bes.: Johannes Geismann
Stab I./KG 77
Mittelmeerraum 1942
Schiffsversenkungen
+ 1 Luftsieg 4. 10. 1942

Ju 88A
Bes.:
Einheit:
Ort:
1944
Schiffsversenkungen

Ju 88A Herbst 1941?
Bes.: Werner Baumbach?
Ort:
Schiffsversenkungen

Bf 109F W. Nr. 109 6469
Frank Liesendahl
6. (Jabo)/JG 2
Cherbourg/Frankr.
9. Juni 1941
Schiffsversenkungen
(6922 Brt.)

Bf 109F W. Nr. 7629
Frank Liesendahl
St. Kap. 10. (Jabo)/JG 2
Beaumont/Frankr.
31. März 1942
Schiffsversenkungen

Handley-Page »Halifax«
mit 12 Raid-Markierungen
im Luftkampf abgeschossen von Werner Streib am 3. 5. 1943

Boeing B-17F -130-BO Serial 230713
Crewchief: Sgt. Blumburg
mit Markierung von 25 Missions und 7 Abschüssen

3. Individual and Humoristic Markings

Since the previous chapters dealt with the more or less official rudder markings a separate chapter on the strictly unconventional ones seems to be in order.

In viewing these pictures the widespread opinion that the German military service could not be surpassed in its stubbornness and single-mindedness may perhaps change slightly. Going »strictly by the book« may have been the case during times of peace and during training but during the war and then especially with the operational units these strict rules were certainly loosened even at the risk of having somebody once in a while going way out. A good example for things like that was offered by the fuselage illustration of a Do 17P during the Polish campaign, showing the pilot as a horse-and-buggy driver »wettening« his »carriage« in accordance with an old custom of that profession. But the very first humoristic illustrations on a German aircraft were very likely borne by Hellmuth Hirth's »Rumpler-Taube« on whose fuselage sides the Munich painter Kneiss had painted some typical Bavarian characters prior to Hirth's participation in the »Kathreiner-Preis« (Munich-Berlin, 29/30 June 1911). During the 1st World War humoristic illustrations on warplanes of all kinds became quite popular with German as well as with Allied pilots.

The years between the wars had military aircraft painted again in their sober, strictly functional schemes, often provided with squadron markings but without any additional extravagant illustrations.

Only with the »Legion Condor« operation in the Spanish Civil War from 1936 on the crews revived the old World War 1 tradition in giving their aircraft some »special« type of markings. In addition to the vertical tailplanes the fuselage sides too served as playgrounds for artistics ambitions. Since, however, this book merely deals with rudder markings these quite often rather picturesque markings had to be left out.

Following the outbreak of World War 2 some Luftwaffe units were again practicing the art of painting rudders and fuselages as sparetime hobbies whereby as a result of the over-

3. Individuelle und humoristische Markierungen

Nachdem die vorangegangenen Abschnitte die mehr oder weniger offiziellen Rudermarkierungen zum Thema hatten, sei nunmehr in diesem Kapitel von den absolut unkonventionellen Bemalungen der Flugzeuge die Rede.

Etwas wird vielleicht beim Betrachten dieser Bilder auch die allgemeine Meinung, der deutsche »Kommiß« sei an Sturheit und tierischem Ernst nicht zu übertreffen gewesen, abgeschwächt. Dieses »strikte nach Vorschrift« mag allenfalls zu Friedenszeiten und während der Ausbildung volle Gültigkeit gehabt haben, im Krieg und da besonders bei den Frontverbänden wußte man »von oben« sehr wohl Zugeständnisse zu machen, selbst auf die Gefahr hin, daß dabei hin und wieder etwas über die Stränge gehauen würde. Ein gutes Beispiel hierfür bietet die Rumpfbemalung einer Do 17P im Polenfeldzug, die einen Flugzeugführer, also einen »Kutscher«, darstellt, der nach althergebrachter Sitte dieses Berufsstandes respektlos sein Flugzeug »befeuchtet«.

Die überhaupt erste humoristische Bemalung auf einem deutschen Flugzeug dürfte die »Rumpler-Taube« von Hellmuth Hirth getragen haben, auf deren Rumpfseiten der Kunstmaler Kneiß aus München vor dem Start zum Flug um den »Kathreiner-Preis« (München–Berlin, 29./30. Juni 1911) urbayrische Typen aufpinselte. Im I. Weltkrieg wurden dann die humorvollen Bemalungen von Kriegsflugzeugen aller Typen auf deutscher Seite, wie auch beim Gegner Legion.

Die Jahre zwischen den beiden Weltkriegen sehen dann die Militärflugzeuge wieder in ihrem nüchternen, rein funktionellen Anstrich, zwar vielfach mit Staffelzeichen versehen, jedoch ohne jegliche extravagante zusätzliche Markierung.

Erst mit dem Einsatz der »Legion Condor« im Spanischen Bürgerkrieg ab 1936, lassen die dort fliegenden Besatzungen die alte Weltkriegstradition wieder aufleben, ihren Maschinen eine Bemalung »mit Pfiff« zu geben. Dabei werden neben den Flächen des Seitenleitwerkes auch die Rumpfseiten zum Austoben der künstlerischen Ambitionen mit herangezogen. Da in diesem Buch jedoch lediglich Leitwerksmarkierungen behandelt werden sollen, müssen diese oft recht originellen Rumpfbemalungen leider unberücksichtigt bleiben.

Auch mit Beginn des II. Weltkrieges wird bei einzelnen Luftwaffeneinheiten als Freizeitbeschäftigung das Bepinseln von Rudern und Rümpfen mit lustigen Emblemen und Szenen praktiziert, dabei

enthusiastic reaction to the first victories the boundaries of good taste and fairness towards the adversary were sometimes crossed. But such things remained isolated instances and were no longer seen in the second year of the war. During the course of the war such additional markings were used less and less and finally disappeared completely. In some instances squadron or group emblems made their appearance on the rudder too but these markings too remained exceptions.

The following pictures show in conclusion some »special« rudder markings that in most cases were recorded in the years 1939 and 1940 when the war situation still allowed German pilots a humoristic point of view.

im Überschwang der Begeisterung über die ersten Erfolge wohl auch manchmal die Grenze des guten Geschmacks über-, die Fairness dem Gegner gegenüber unterschritten. Jedoch bleiben solche Entgleisungen Einzelerscheinungen, und sind bereits im zweiten Kriegsjahr nicht mehr anzutreffen. Mit der weiteren Entwicklung der kriegerischen Ereignisse werden naturgemäß solche zusätzlichen Bereicherungen des Flugzeuganstriches immer seltener, und verschwinden schließlich ganz. In wenigen Fällen erscheinen auch Staffel- oder Gruppenzeichen am Seitenleitwerk, bilden jedoch Ausnahmen.

Die nachstehenden Photos zeigen abschließend einige Ruderbemalungen »außer der Reihe«, mit wenigen Ausnahmen in den Jahren 1939 und 1940 aufgenommen, als der Kriegsverlauf deutscherseits noch humoristische Betrachtungen zuließ.

Rumpler »Taube«
Hellmuth Hirth, München 29. Juni 1911
Flug um den »Kathreiner-Preis«

He 111B-1
Bes.:
K 88 der »Legion Condor«
Spanien

He 111B-1
Bes.:
Zaragosa/Spanien

der K 88 »Legion Condor«
1938

Ein Kuriosum: Bei der K 88 der »Legion Condor« hat es das Maskottchen, den Terrier »Peter«, der alle Einsätze mitflog über Sagunta bei einem Luftkampf erwischt. Die Besatzung vermerkt den »Trauerfall« auf dem Ruder ihrer He 111 B.

A curiosity: With the K 88 of the »Legion Condor« was as a mascot the scotch terrier »Peter« that flew on all missions until being killed in a dogfight over Sagunta, the crew marking his sad ending on the stabilizer of their He 111 B.

Do 17P
Besatz.:
Polenfeldzug 1939

Einh.:

He 111P
Besatz.:
Winter 1939/40

I./KG 27

He 111P
Bes.:
Delmenhorst

III./KG 27
1939

He 111P
Bes.:
1939/40

W. Nr. 1614
Einheit:

He 111P
Bes.:
KG 27
Winter 1939 40

He 111P W. Nr. 1992
Bes.:
KG 27
Winter 1939/40

He 111P
Bes.:
KG 27
Winter 1939/40

He 111P W. Nr. 1524
Bes.:
KG 27
Winter 1939/40

He 111P
Bes.:
I./KG 27
Winter 1939/40

He 111P W. Nr. 1992
Bes.:
KG 27
Winter 1939/40

He 111P
Bes.:
KG 27
Winter 1939/40

He 111H
Bes.: Franz Schmitt
St. Kap. 14. (Eis)/KG 55
Ostfront

He 111P W. Nr. 1485
Bes.:
KG 27
Winter 1939/40

Ju 87D
Bernhard Wuttka
Ostfront

St. Kap. 8./St. G. 2
Frühjahr 1943

Ebersberger, Kurt
4./JG 26
FW 190 A
Abbeville-Drucat
9. Mai 1942
23. L.S. (27)

Gollob, Gordon
Kdore. JG 77
Bf 109 F
Oktoberfeld/Krim
21. Juni 1942
107. L.S. (150)

Graf, Hermann Kertsch/Krim 104. L.S. (212)	Staffkap. 9./JG 52 14. Mai 1942	Bf 109 F

Trautloft, Johannes Siwerskaja/Ostfr. 26. L.S. (57)	Kdore. JG 54 Winter 1941/42	Bf 109 F

Weißenberger, Theo
St.Kap. 7./JG 5
Bf 109 G
Petsamo/Finnl.
25. Juli 1943
112. L.S. (208)

Die Me 262A, die Theo Weißenberger als Kommandeur der I./JG 7 im Herbst 1944 flog.

When Kommandeur of I. Gruppe/JG 7 Theo Weissenberger piloted this Me 262A in Autumn 1944.

Schroer, Werner 2./JG 27 Bf 109 E
Nordafrika Sept. 1941
4. L.S. (114)

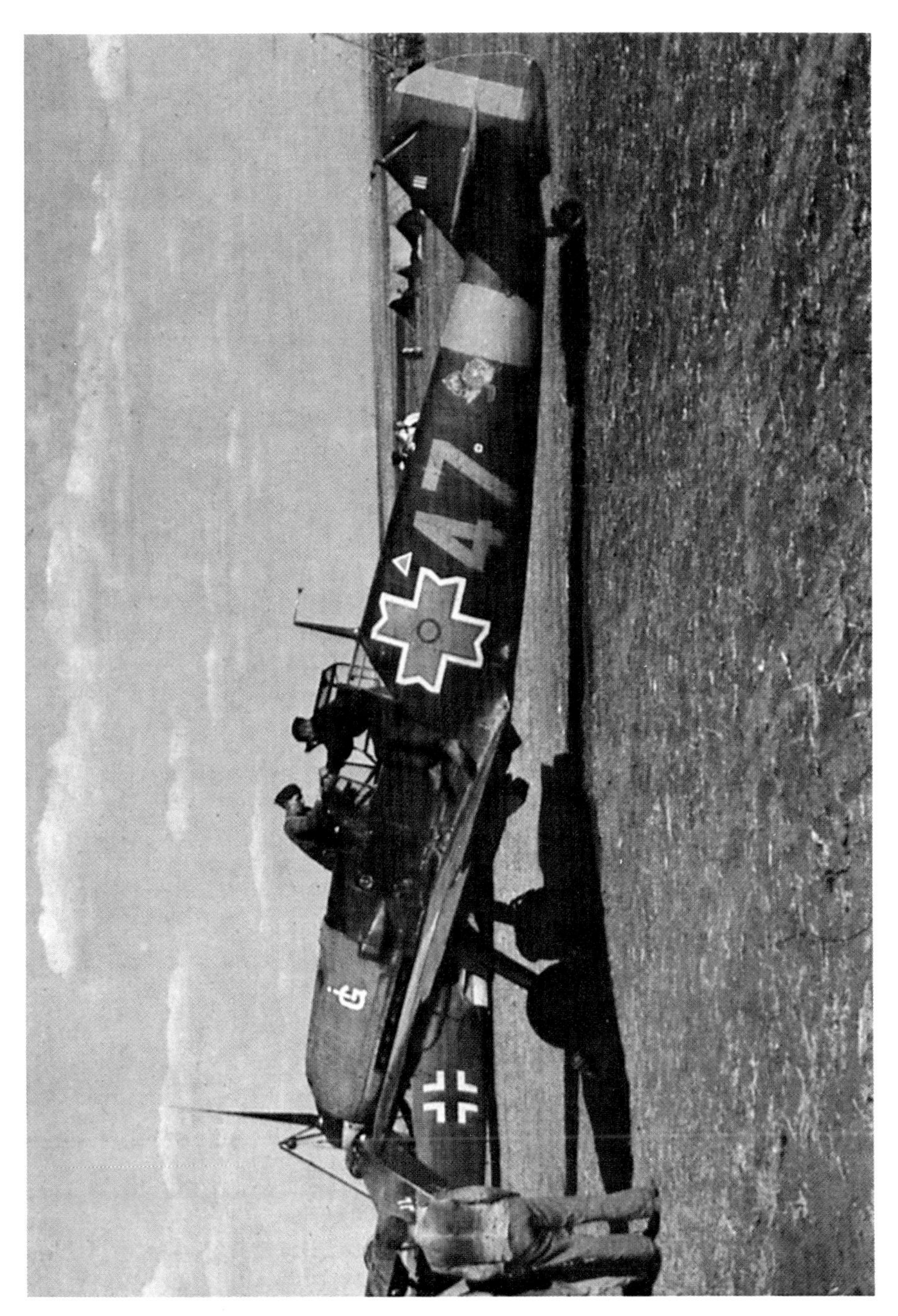

Bf 109 E einer rumänischen Jagdstaffel. Sommer 1941
Jassi/Rum.
3. L.S.

Bf 109 F
Ostfront
Frühjahr 1942

FW 200 C
KG 40
Südfrankr.

Namensregister

Ehemalige Luftwaffenangehörige werden darum gebeten, weitere, hier noch nicht erfaßte Ruderaufnahmen den Autoren leihweise zur Verfügung zu stellen, um durch einen Ergänzungsband das Sachgebiet Abschußmarkierungen vervollständigen zu können.

K. Ries
65 Mainz-Finthen
Postfach 25

E. Obermaier
8 München
Implerstraße 46/V

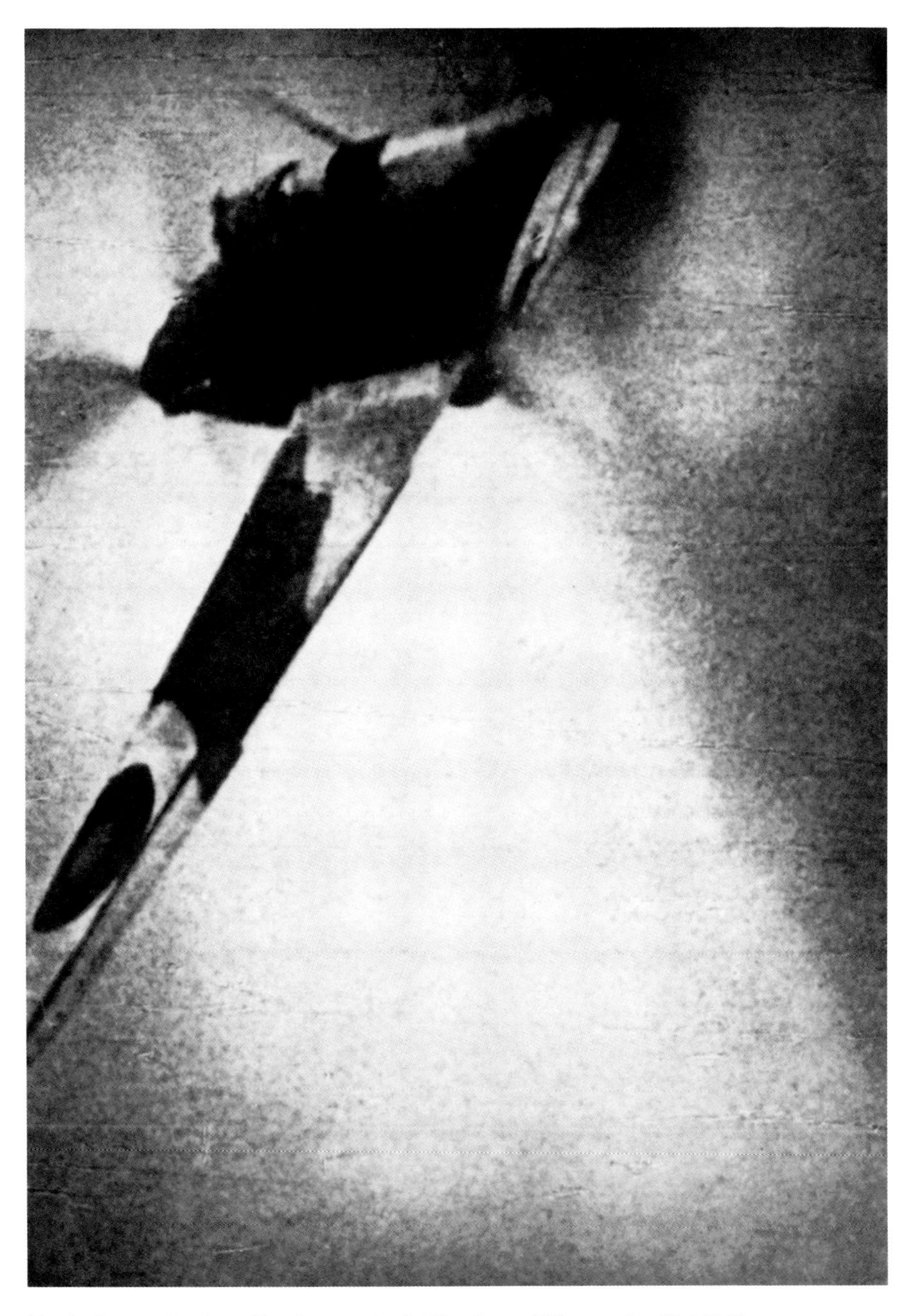

Abschuß einer Hawker »Hurricane« durch Ofw. Franz **Götz** von der III./JG 53.
Gun-camera picture of a Hawker »Hurricane« deadly hit by Ofw. Franz **Götz,** III./JG 53.

Überschlag eeiner sowjetischen R-5, die mit starken Beschußschäden eine Notlandung in der Nähe von Smolensk versuchte.

After a forced landing near Smolensk/Russia this R-5 did a ground loop. Both crew members escaped unhurt.

Maj. Heinz **Bär** an der Notlandestelle der von ihm abgeschossenen Boeing B-17F-70-BO Serial 23040. Rechts sein Rottenflieger Leo Schuhmacher.

Maj. Heinz **Bär** inspects the B-17F-70-BO Serial 23040 brought down by him over German territory. At the far right his wing man Leo Schuhmacher.

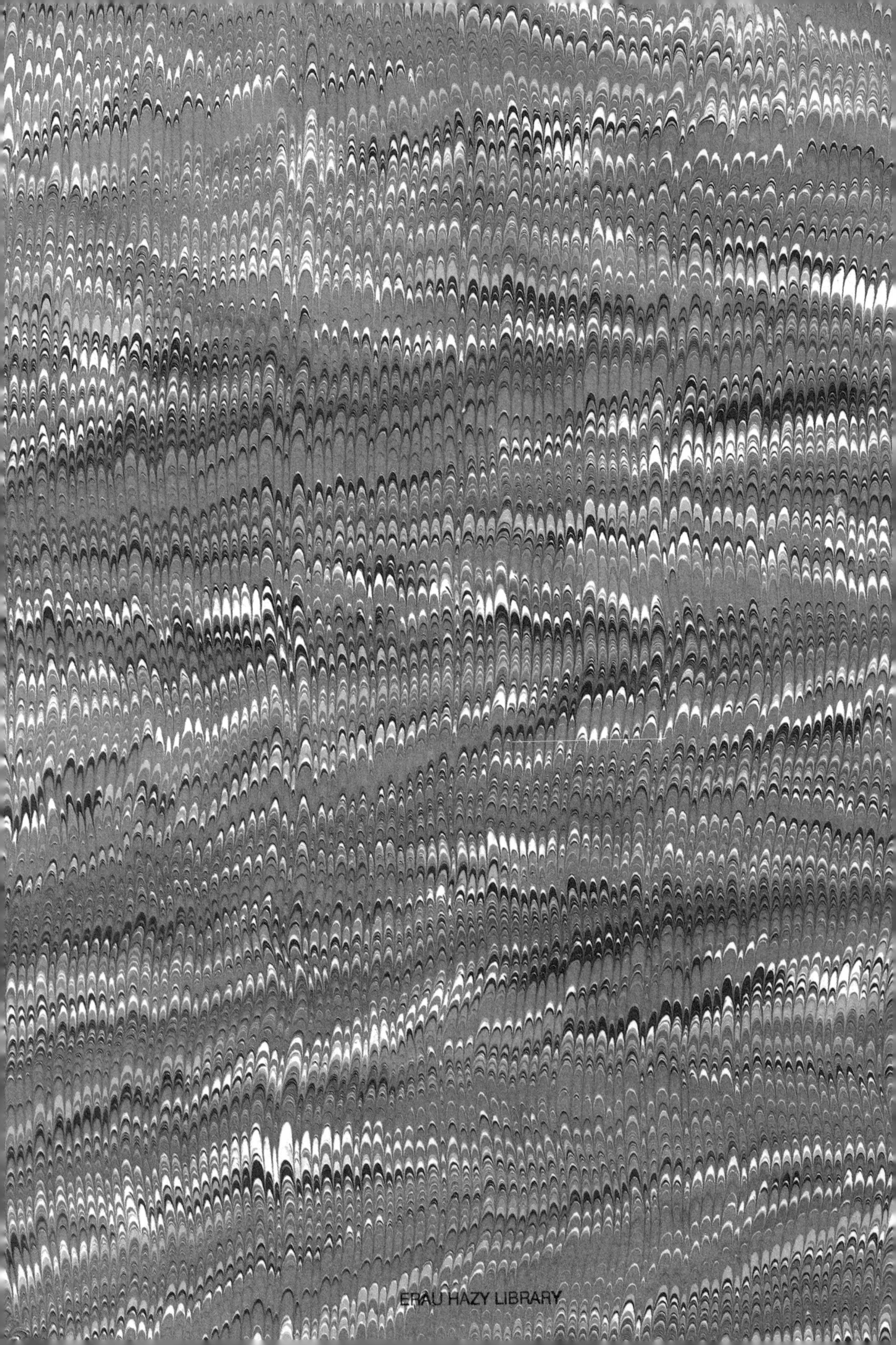